Hanns Fischer   Heinz Heimgartner (Eds.)

Organic Free Radicals

Proceedings of the Fifth International Symposium
Zürich, 18.–23. September 1988

Springer-Verlag
Berlin Heidelberg New York
London Paris Tokyo

Prof. Dr. Hanns Fischer
Physikalisch-Chemisches Institut
Universität Zürich
Winterthurerstraße 190
8057 Zürich, Switzerland

Prof. Dr. Heinz Heimgartner
Organisch-Chemisches Institut
Universität Zürich
Winterthurerstraße 190
8057 Zürich, Switzerland

ISBN 3-540-50129-0 Springer-Verlag Berlin Heidelberg New York
ISBN 0-387-50129-0 Springer-Verlag New York Berlin Heidelberg

Printing and binding: Druckhaus Beltz, 6944 Hemsbach
2151/3140-543210 – Printed on acid-free paper

F O R E W O R D

The "5th International Symposium on Organic Free Radicals",
jointly organized by Chemistry Institutes of ETH and University
of Zürich, will be held at the Irchel campus of the University
from September 18 to 23, 1988. About 200 participants from all
over the world will come together for this symposium to exchange
and discuss scientific results, to get in personal contact with
scientists from other countries, and to meet collegues and
friends. We hope that besides the scientific presentations there
will be the opportunity for many formal and informal discussions.

This volume contains the abstracts received up to June 13 of the
planned fifteen plenary lectures, nine invited talks, and about
ninetyfive contributed papers which will be given as posters. The
abstracts have been reproduced as submitted by the authors, they
are grouped in alphabetical order using the name of the author
giving the lecture or presenting the poster. In addition, an
author and a subject index facilitate the finding of
contributions. We are grateful to all of the participants who have
submitted abstracts and we note that the wide range of topics
covered in these abstracts - ranging from theoretical aspects and
spectroscopy of radicals, reaction mechanisms and synthetic
applications, to the significance of radicals in biological and
medical chemistry - indicates the developement and importance of
chemistry with radicals.

This meeting follows previous symposia held in Sirmione (Italy)
1974, Aix-en-Provence (France) 1977, Freiburg i.Br. (Federal
Republic of Germany) 1981, and St. Andrews (Great Britain) 1984.
The program of this 5th Symposium consists of plenary lectures in
the mornings and three poster sessions and Minisymposia in the
afternoons. The topics of the Minisymposia, indicating three main
fields of this meeting, are *Free Radicals in Organic Synthesis*
(Chairman *B. Giese*), *Free Radicals in Biological Systems*
(Chairman *W. Adam*), and *Electron Transfer in Free Radical
Chemistry* (Chairman *L. Eberson*).

The Organizing Committee is pleased to acknowledge financial support by *ETH-Zürich, Kanton Zürich, Universität Zürich, Zürcher Hochschulverein, Schweizerischer Nationalfonds zur Förderung der wissenschaftlichen Forschung, Schweizerische Naturforschende Gesellschaft, Ciba-Geigy AG, Givaudan Forschungsgesellschaft AG, F. Hoffmann-La Roche & Co. AG, Lonza AG, Sandoz AG, Bruker AG Karlsruhe, Spectrospin AG Schwerzenbach, Varian AG Schweiz,* and *Springer-Verlag GmbH & Co. AG, Heidelberg.* The realization of the symposium was made possible by the generosity of all these donors.

The International Advisory Committee (*Sir D.H.R. Barton, A.L.J. Beckwith, K.U. Ingold, J.K. Kochi, F. Minisci, N. Ono, Ch. Rüchardt, J.M. Surzur, G. Stork, Lord J.M. Tedder,* and *H.G. Viehe*) and the Local Organizers (*D. Bellus, H. Fischer* (Chairman), *H.-J. Hansen, H. Heimgartner, B. Kräutler, H. Paul, D. Seebach, A. Vasella,* and *W.-D. Woggon*) are to be thanked for their assistance in arranging the technical program and in making all of the local arrangements. Particular thanks are due to *Dr. R. Stumpe*, Springer Verlag, for his efforts in printing of these Proceedings.

We look forward to a fruitful scientific meeting which not only will help to improve our knowledge on free radicals in chemistry but also will stimulate new work and collaboration in this field. Last but not least we hope that the Symposium will promote human contacts between scientists from different countries.

Zürich, June 15, 1988   Hanns Fischer
                                         Heinz Heimgartner

Table of Contents

THERMAL DECOMPOSITION OF INDOLINONIC NITROXIDES.

A KINETIC ESR STUDY

A. ALBERTI,[a] P. CARLONI,[b] L. GRECI,[b] and P. STIPA[b]

[a] I. Co. C. E. A. - C. N. R., I-40064 OZZANO EMILIA, Italy
[b] Facoltà di Ingegneria, Università, I-60131 ANCONA, Italy

Indolinonic nitroxides (1, X = O, NPh) may be viewed as the cyclic analogues of alkyl aryl nitroxides, but are characterized by a persistence comparable to that of tetramethyl-pyrrolidino- and tetramethyl-pyperidino-1-oxyls.

Although solutions of aminoxyls (1) can be kept for a long time at room temperature without showing any appreciable modification of the intensity of their ESR signals, these radicals may undergo thermal degradation to the N-oxide (2) and its deoxygenated analogue (3), the reaction proceeding *via* initial loss of the ·R radical.

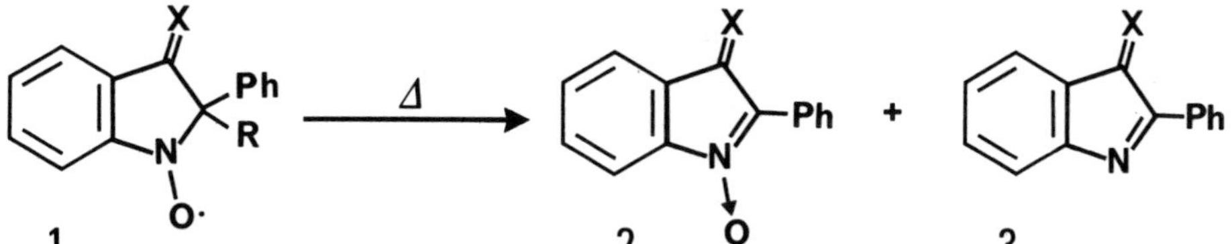

X = O, NPh	a: R = Et	d: R = Benzyl
	b: R = ^{i}Pr	e: R = Allyl
	c: R = ^{t}Bu	

The disappearence of the ESR signal of various nitroxides (1) with time was found to follow clean first order kinetics, thus allowing the determination of the rate constants for the degradation processes. The activation parameters could be derived only for (1, X = NPh R = benzyl, allyl, and ^{t}Bu); however at 200°C the rate constants

H. Fischer, H. Heimgartner (Eds.)
Organic Free Radicals
© Springer-Verlag Berlin Heidelberg 1988

K_1 increased in the order Et < i-Pr < allyl < benzyl < t-butyl.

In principle one should expect the existence of a linear correlation between the rate constants k_1 and the stability of the radicals ·R, and hence the bond dissociation energy (BDE) of the R-H bond in the corresponding hydrocarbons. A linear correlation was in fact found to hold for nitroxides 1 (X = NPh), with the exception of the ^{t}butyl substituted compound, whose degradation was much faster than expected. The different factors which may possibly originate this odd behaviour are examined.

RADICAL CATIONS

FROM ALKYL IODIDES AND 1,n-HALOGEN(ALKYLTHIO)ALKANES
IN SOLUTION

K.-D. Asmus, E. Anklam and H. Mohan

Hahn-Meitner-Institut Berlin, Bereich Strahlenchemie
Postfach 39 01 28, D-1000 Berlin 39

Abstract: Radical cations of the general type $(\text{-I}\therefore\text{I-})^+$, $(\text{=S}\therefore\text{I-})^+$, $(\text{=S}\therefore\text{Br-})^+$, and $(\text{=S}\therefore\text{S=})^+$ with optical absorptions in the UV/VIS are formed upon one-electron oxidation of alkyl iodides and 1,n-halogen(alkylthio)alkanes.

RADICAL CATIONS FROM ALKYLIODIDES AND DIIODOALKANES

Radical cations of type (I) and (II) with a $2\sigma/1\sigma^*$ three-electron bond between two iodine atoms can be generated upon ·OH radical induced oxidation of alkyliodides and various 1,n-diiodoalkanes in acid (pH < 5) aqueous solutions [1,2]. The formation of intermolecular radical cations (I)

$$(\text{R-I} \therefore \text{I-R})^+ \qquad\qquad \left[\begin{array}{c} \text{I} \therefore \text{I} \\ \text{(CH}_2)_n \end{array} \right]^+$$

(I)   (II)

requires relatively high RI concentrations with maximum yields being achieved at $>5\times10^{-3}$ M. Stabilization of (II) occurs independent of solute concentration by intramolecular interaction of the two iodine atoms in the oxidized diiodo compound. Electron release by the substituent R into the three-electron bond leads to bond weakening and is reflected in a corresponding red-shift of the optical absorption. The respective λ_{max} of (I) range from 415 nm for R=Me to 470 nm for R= s-but [1].

The stability of the intramolecular species (II) is mainly controlled by steric parameters. The most favourable structure is achieved if the radical cation assumes a five-membered ring configuration which is the case for n=3. In going from n=3 to n=6 the λ_{max} of (II) undergo a red-shift from 405 - 455 nm while the respective half-lives of the radicals decrease from 150μs to 75 μs [2].

H. Fischer, H. Heimgartner (Eds.)
Organic Free Radicals
© Springer-Verlag Berlin Heidelberg 1988

The ·OH induced oxidation generally proceeds via an adduct intermediate. This undergoes a proton assisted dissociation into the molecular radical cation prior to the establishment of the three-electron bonded system, e.g.

$$RI + \cdot OH \longrightarrow RI(OH)\cdot \xrightarrow{\quad H^+ \quad} RI^{\cdot +} \xrightarrow{\quad RI \quad} (RI \cdot \cdot IR)^+$$

All three oxidized forms of the alkyl iodide are highly oxidizing species. The $CH_3I^{\cdot +}$, for example, has a redox potential of at least +2.0 V (vs. NHE) and readily oxidizes, among others, metal ions like Ag^+ and organic sulfides with diffusion controlled rate constants [3].

RADICAL CATIONS FROM 1,n-HALOGEN(ALKYLTHIO)ALKANES

One-electron oxidation of 1,n-halogen(alkylthio)alkanes leads to intra- as well as intermolecular radical cations depending on the nature of the halogen and the solute concentration [4,5].

$$\left[R-S \cdot \cdot Hal \right]^+ \qquad \left[\begin{array}{c} Hal-(CH_2)_n \\ \diagdown \\ S \cdot \cdot Hal-(CH_2)_n-S-R \\ \diagup \\ R \end{array} \right]^+ \qquad \left[\begin{array}{c} R \qquad\qquad R \\ \diagdown \qquad \diagup \\ S \cdot \cdot S \\ \diagup \qquad \diagdown \\ Hal(CH_2)_n \quad (CH_2)_nHal \end{array} \right]^+$$

(III)   (IV)   (V)

Intramolecular radical cations of type (III) have been observed for iodine and bromine compounds at low solute concentrations. At higher solute concentrations (typically > 10^{-3} M) intermolecular sulfur-halogen species (IV)(with Hal = I in non-aqueous solutions) and purely sulfur centered species (V) are generated. The latter are the only type of stabilized radical cations in case of chloro compounds.

All species (III)-(V) exhibit optical absorptions in the UV/VIS with similar characteristics and dependences as mentioned above for the iodine centered radical cations and as described in detail in our earlier work on the oxidation of unsubstituted organic sulfides [6].

1 Mohan H, Asmus K-D (1987) J Chem Soc Perkin Trans 2 :1795
2 Mohan H, Asmus K-D (1987) J Amer Chem Soc 109:4745
3 Mohan H, Asmus K-D (1988) J Phys Chem 92:118
4 Anklam E, Mohan H, Asmus K-D (1987) Helv Chim Acta 70:2110
5 Anklam E, Mohan H, Asmus K-D (1988) J Chem Soc Perkin Trans 2 : 000
6 Asmus K-D (1979) Acc Chem Res 12:436

STUDIES OF ß-SULFONYL SUBSTITUTED VINYL NITROXIDES

H.G. Aurich and K.-D. Möbus

Fachbereich Chemie, University of Marburg
Hans-Meerwein-Strasse, D-3550 Marburg, FRG

Vinyl nitroxides **3** could be detected when hydroxylamines **1** were oxidized by lead dioxide. The oxidation proceeds via the corresponding t-butyl alkylnitroxides and the nitrones **2**. In certain cases spin adducts formed by addition of vinyl nitroxides **3** to nitrones **2** were also observed.

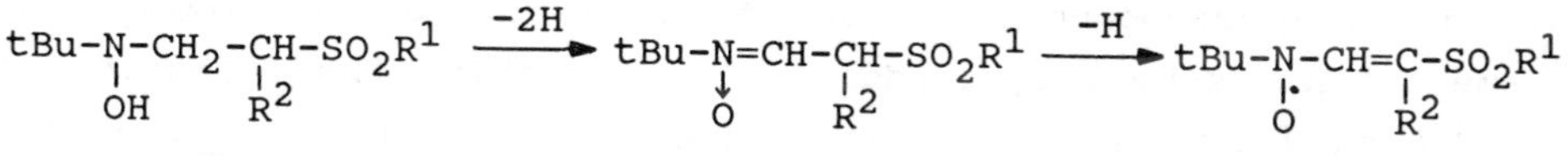

$$tBu\underset{\underset{OH}{|}}{N}-CH_2-\underset{\underset{R^2}{|}}{C}H-SO_2R^1 \xrightarrow{-2H} tBu\underset{\underset{O}{|\cdot}}{N}=CH-\underset{\underset{R^2}{|}}{C}H-SO_2R^1 \xrightarrow{-H} tBu\underset{\underset{O}{|\cdot}}{N}-CH=\underset{\underset{R^2}{|}}{C}-SO_2R^1$$

$$\textbf{1} \qquad\qquad\qquad \textbf{2} \qquad\qquad\qquad \textbf{3}$$

$$tBu\underset{\underset{O}{|\cdot}}{N}-CH=\underset{\underset{SO_2tBu}{|}}{C}----\underset{\underset{O}{\downarrow}}{C}H=N-tBu \longleftrightarrow tBu\underset{\underset{O}{\downarrow}}{N}=CH-\underset{\underset{SO_2tBu}{|}}{C}=\!=\!=CH-\underset{\underset{O}{|\cdot}}{N}-tBu \qquad \textbf{4}$$

ESR coupling constants (in Gauß) of the vinyl nitroxides **3** and **4** in $CHCl_3$

	R^1	R^2	a^N	$a^H_{\alpha-C}$	$a^H_{R^2}$	Temp.[°C]
3a	$C_6H_4-CH_3-4$	Me	8.65	2.5	8.55(3H)	+20
3b	t-Bu	$CH_2-CH=CMe_2$	8.25	2.65	9.7(1H),<0.3(1H)	+20
			8.25	2.65	4.9(2H)	+60
3c	C_6H_5	H	8.2	1.65	7.3(1H)	+20
3d	$CH_2-CH=CH_2$[a]	H	8.0	1.5	8.0(1H)	+20
4	-	-	4.75(1N) 3.3 (1N)	4.75(1H) 3.3 (1H)		-38
			4.03(2N)	4.03(2H)		+65

[a] an additional coupling of three protons of the allyl group is found: $a^H = 1.1$ G (3H)

H. Fischer, H. Heimgartner (Eds.)
Organic Free Radicals
© Springer-Verlag Berlin Heidelberg 1988

In radical **4** delocalization of the unpaired electron through the entire π-system occurs. Since the values of a^N as well as those of a^H are different at $-38°C$ the sterically crowded molecule must exist in an asymmetric form. However, an interconversion of two equivalent conformations of the molecule occurs as is indicated by the ESR spectrum at $65°C$.

Dimerization of radical **4** yields the C.C-bonded dimer **5** (R^1 = tBu, R^2 = CH=N(O)tBu). Compound **6** (R^1 = Ph) is formed from **3c** via **5** (R^1 = Ph, R^2 = H) revealing that C.C-dimerization followed by elimination of benzene sulfinic acid is the main reaction pathway of vinyl nitroxide **3c**. The same reaction sequence occurs with vinyl nitroxide **3d**, however, compound **7** formed in this way undergoes rapidly an intramolecular 1.3-diploar cycloaddition to give bicyclic compound **8**. In addition, compound **10** arises from the O.C-bonded dimer **9** by the same intramolecular cycloaddition. Presumably, the formation of such O.C-bonded dimers is generally reversible so that this reaction channel is only productive if an irreversible reaction step follows, as for instance **9** → **10**.

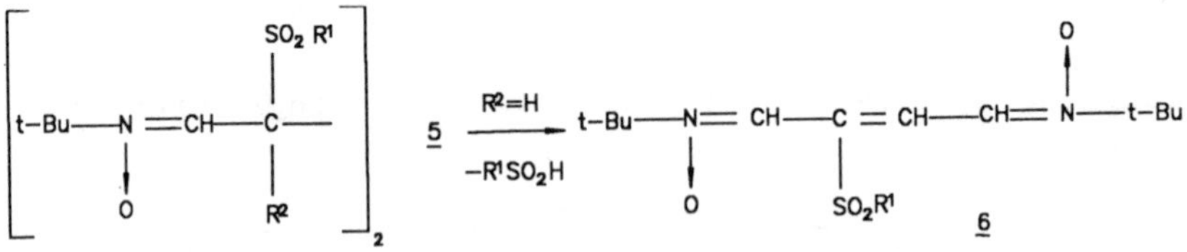

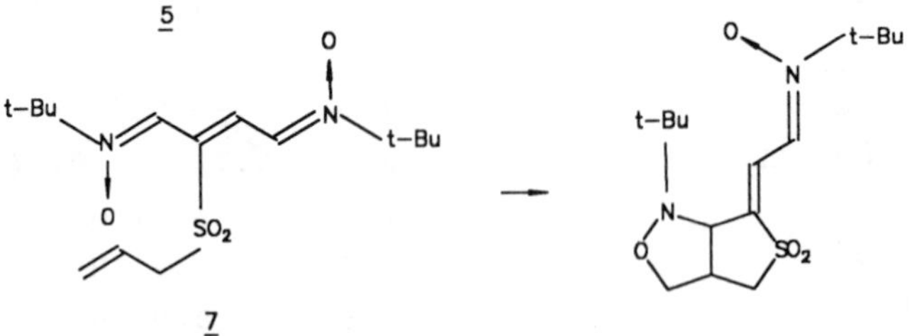

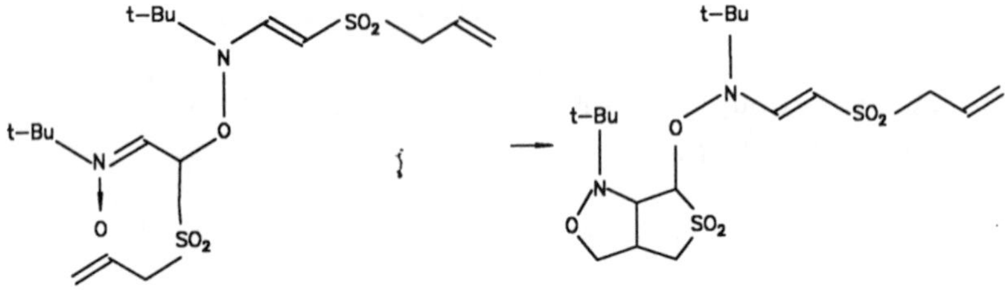

DIRECT AND INDIRECT DIELS-ALDER ADDITIONS
TO ELECTRON RICH DIENOPHILES

N. L. Bauld

Department of Chemistry, The University of Texas,
Austin, Texas 78712 USA

Abstract: Direct and indirect cation radical Diels-Alder cycloadditions of conjugated dienes to electron rich alkenes, including
vinyl ethers, vinyl sulfides, and N-vinylamides, are developed and
applied to the synthesis of the sesquiterpenoid natural product
β-selinene.

The intermolecular Diels-Alder cycloaddition of electron rich
alkenes to conjugated dienes is a potentially valuable synthetic
operation which has been mechanistically blocked by the low cycloaddition reactivity of this sub-class of dienophiles. Recent
observations of cation radical Diels-Alder cycloadditions of vinyl
ethers and vinyl sulfides to conjugated dienes suggests a novel
catalytic approach to this challenging methodological problem. The
feasibility and synthetic utility of the cation radical DA strategy
for cycloadditions to electron rich alkenes is underscored by the
synthesis of the sesquiterpenoid natural product β-selinene using
an efficient cation radical DA cycloaddition to phenyl vinyl sulfide
(PVS). In sharp contrast to the efficiency and selectivity of the
cation radical addition to PVS, the corresponding thermal addition
to phenyl vinyl sulfone is grossly unselective and inefficient.
The fact that Diels-Alder (DA) periselectivity in cation radical
cycloadditions of electron rich alkenes to conjugated dienes is often
not observed when the conjugated diene exists predominantly in the
s-trans conformation significantly limits the above strategy, however.
Consequently, a general strategy for indirect DA cycloaddition to
electron rich alkenes has been developed for such dienes. This
strategy involves initial cation radical cyclobutanation of a
predominantly s-trans diene, followed by a vinylcyclobutane rearrangement to the net DA adducts. Two discrete versions of this latter
rearrangement have been developed. The first involves an alpha
heteroatom anion assisted VCB rearrangement, which is well known in
the oxy anion case, but which is established for the first time in the

H. Fischer, H. Heimgartner (Eds.)
Organic Free Radicals
© Springer-Verlag Berlin Heidelberg 1988

nitrogen anion case in the present work. The second version,
appropriately, is a novel cation radical VCB rearrangement, which
emerges as an incredibly facile process especially when ionizable
substituents (anisyl, phenylthio, phenoxy) are present at the
2-position of the VCB. However, VCB rearrangements of even
hydrocarbon VCB's are observed, and the stereochemistry of the
rearrangement is explored. Appropriate control experiments establish
the intramolecular nature of the VCB‡ rearrangement, and sr
stereochemistry is observed in several instances.

Cycloadditions of dienes/electron rich alkenes appear to
proceed predominantly _via_ a gross mechanism involving the diene
cation radical. The reactivity towards such diene cation radicals,
called caticophilicity by this group, is at its highest level for
enamides and vinyl sulfides.

PHOTO-FRIEDEL-CRAFTS REACTION IN SYNTHESIS
OF CATHARANTHINE AND ITS DERIVATIVES

H. Bölcskei[a], E. Gács-Baitz[b] and Cs. Szántay[b]

a) Chemical Works of Gedeon Richter Ltd., H-1475 Budapest, POB 27
b) Central Research Institute for Chemistry, Hung. Acad. Sci.,
   H-1525 Budapest, POB 17 (Hungary)

Abstract: Catharanthine 1a, allocatharanthine 2a and 20-deethyl-
catharanthine 3a were synthetised using a photochemical key step.

The Catharanthus roseus alkaloids vincristine and vinblastine are
widely used in the chemotherapy of cancer. Catharanthine 1a is a
major alkaloid of the same plant. On coupling catharanthine or its
derivatives (allocatharanthine 2a, 20-deethylcatharanthine 3a) with
vindoline vinblastine type diindole compounds can be obtained [1,2]
which have antitumor activity.

In the course of our synthesis of (+)-20-deethylcatharanthine 3a [3]
precursors 6a and 6b were prepared and their ring closure reactions
leading to the Iboga-type skeleton were also studied. (For preparation
of 6a and 6b see [3]). Heating of 6a resulted in new rearrangement
products [4]. 6b did not react in presence of Friedel-Crafts catalysts
($AlCl_3$, $SnCl_4$, $TiCl_4$, etc.). Only the photocyclisation was successful.
Irradiation of N-(indolyl-acetyl)-isoquinuclidine in methanolic
solution 6b resulted in three main products: 3b the desired 5-oxo-20-
-deethylcatharanthine 16-20 %, the known intermediate of 3a [5], its
regioisomer 7 and the byproduct 8. To avoid this byproduct several
solvents were tried. Using tetrahydrofurane/water as solvent the ring
closured products 3b and 7 and an uncyclised new rearranged compound
9 having a hydroxi group instead of chlorine atom were obtained.

The N-(indolyl-acetyl)-isoquinuclidine derivatives 4b and 5b were
synthetised on a similar way as 6b. (For preparation of 4b and 5b

H. Fischer, H. Heimgartner (Eds.)
Organic Free Radicals
© Springer-Verlag Berlin Heidelberg 1988

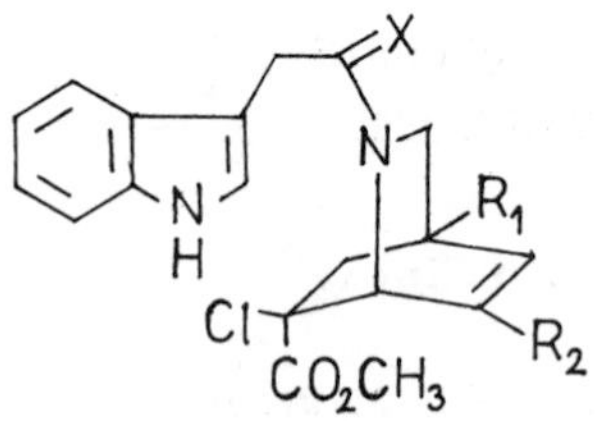

$\underline{4}$ R_1=H, R_2=C_2H_5 $\underline{a}$ series X=H_2

$\underline{5}$ R_1=C_2H_5, R_2=H $\underline{b}$ series X=O

$\underline{6}$ R_1=R_2=H

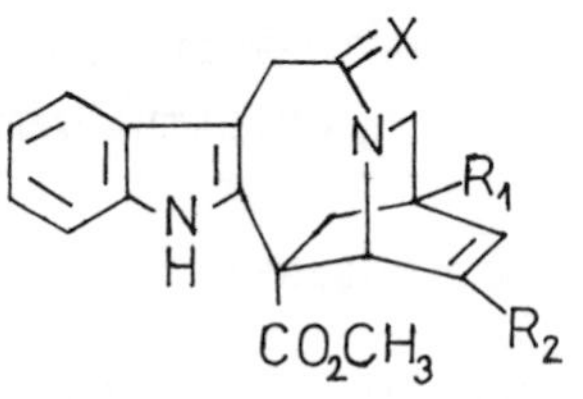

$\underline{1}$ R_1=H, R_2=C_2H_5

$\underline{2}$ R_1=C_2H_5, R_2=H

$\underline{3}$ R_1=R_2=H

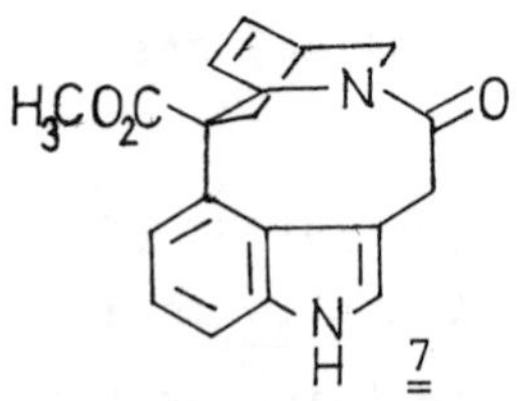

see [6,7]). Fortunately the photochemical reaction of $\underline{4b}$ and $\underline{5b}$, led to the products $\underline{1b}$ and $\underline{2b}$, respectively, in higher yield (~30-30 %). The oxo groups of $\underline{1b}$ and $\underline{2b}$ were removed by a selective way resulting in $\underline{1a}$ and $\underline{2a}$, respectively.

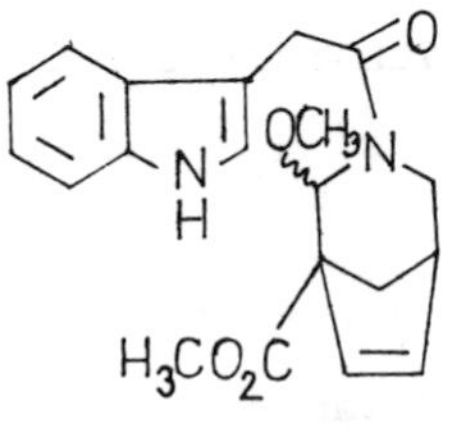

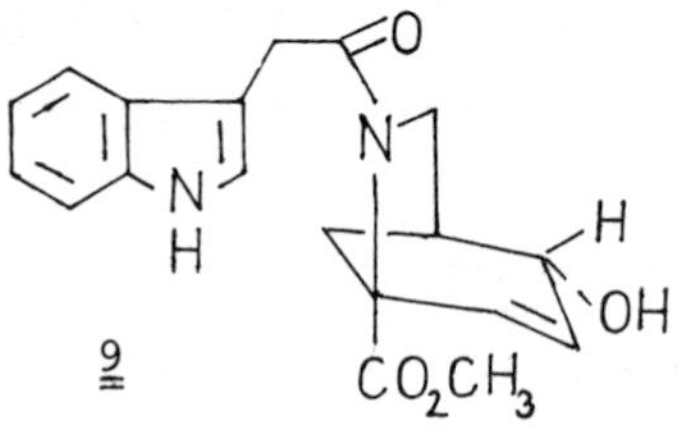

All the spectral data (IR, MS, ^{1}H and ^{13}C NMR) of the new products were in accord with the given structures.

Acknowledgement: We are grateful to József Tamás[b] for mass spectral measurements and to László Párkányi[b] for X-Ray analysis of compound $\underline{7}$.

References:

1 Langlois N, Guéritte F, Langlois Y, Potier P, (1976) J Am Chem Soc 98:7017
2 Guéritte F, Langlois N, Langlois Y, Sundberg RJ, Bloom JD (1981) J Org Chem 46:5393
3 Szántay Cs, Keve T, Bölcskei H, Ács T (1983) Tetrahedron Lett 24:5539
4 Szántay Cs, Keve T, Bölcskei H, Megyeri G, Gács-Baitz E (1985) Heterocycles 23:1885
5 Sundberg RJ, Bloom JD (1980) J Org Chem 45:3382
6 Hung Pat 191 535 (2343/83)
7 Hung Appl 5715/87

HYDROXYL RADICAL-INDUCED STRAND BREAK FORMATION IN SINGLE-STRANDED POLYNUCLEOTIDES AND SINGLE-STRANDED DNA IN AQUEOUS SOLUTION AS MEASURED BY LIGHT SCATTERING AND BY CONDUCTIVITY.

E. Bothe, M. Adinarayana and D. Schulte-Frohlinde

Max-Planck-Institut für Strahlenchemie, Stiftstra-e 34-36,

D-4330 Mülheim a.d. Ruhr

Abstract: Reproductive cell death and loss of biological activity caused by ionising radiation are currently believed to be the result of damage to DNA. Herewith strand breaks in the DNA backbone constitute a quite severe type of damage. Under conditions which favour the indirect effect, the strand breaks are formed predominantly by OH radicals.

In recent studies with the model system poly(U),[1,2] it has been found that with this polymer OH radical-induced strand breakage is much higher than found with DNA. These high yields have been interpreted as indicating a radical transfer reaction from the base moiety to the ribose moiety of poly(U). Such types of reactions evidently occur to a much smaller extent, if at all, in DNA. Therefore it seemed useful to investigate whether the structure, the type of base, the presence of the OH group in the C(2') position or a combination of the above are responsible for the differences in the yields of strand break formation in poly(U) and in DNA.

Combining conductivity measurements and molecular weight determination by means of low-angle laser light scattering, we found for the polyribonucleotides poly(U), poly(A) and poly(C) and for single-stranded DNA that, in presence and absence of oxygen, on average 8.5 counterions per single-strand break are liberated under salt-free conditions. This relationship allowed us to estimate, from conductivity measurements alone, the following G-values of single-strand break formation (in μmol/J) for polydeoxyribonucleotides under anoxic conditions: poly(dA), 0.23; poly(dC), 0.14; poly(dT), 0.06 and poly(dU), 0.046.

By time-resolved conductivity measurements in pulse radiolysis we have measured also the rate of strand break formation. The rate has been found to be similar for poly(dA) and ssDNA over a range of pH values. Poly(dC) and poly(dU) exhibit conductivity increase components with half lives similar to those of poly(dA) and ssDNA at corresponding pH values. The implications of these results are discussed.

References:
1. Lemaire DGE, Bothe E, Schulte-Frohlinde D (1984) Int J Radiat Biol 45:351
2. Lemaire DGE, Bothe E, Schulte-Frohlinde D (1987) Int J Radiat Biol 51:319

H. Fischer, H. Heimgartner (Eds.)
Organic Free Radicals
© Springer-Verlag Berlin Heidelberg 1988

COBALOXIME-MEDIATED RADICAL
CROSS COUPLING REACTIONS

Bruce P. Branchaud,[*1] Mark S. Meier, Youngshin Lee Choi, Gui-Xue Yu

Department of Chemistry, University of Oregon
Eugene, Oregon 97403-1210 USA

Abstract: Studies of novel radical cross coupling reactions using alkyl cobaloximes are reported, including the scope of the reactions, mechanistic aspects, and synthetic applications.

We have discovered and have been developing alkyl cobaloxime mediated radical cross coupling reactions.[2-4] This poster will present our recent studies of three distinct types of cobaloxime-mediated radical cross couplings; alkyl-alkenyl, alkyl-protonated heteroaromatic, and alkyl-nitroalkylanion. From the synthetic perspective, a novel feature of these cobaloxime-mediated radical cross couplings, compared to other synthetically useful radical cross couplings, is the regeneration of the "olefin" functionality in the final product. From a mechanistic perspective, a novel feature is non-chain radical generation and trapping.

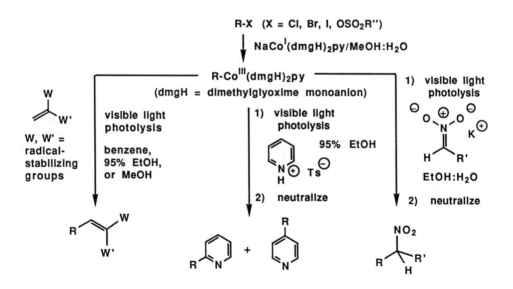

H. Fischer, H. Heimgartner (Eds.)
Organic Free Radicals
© Springer-Verlag Berlin Heidelberg 1988

These reactions can all be rationalized as: photolytic C-Co bond homolysis to produce alkyl radical + $py(dmgH)_2Co^{II}$, alkyl radical addition to the cross coupling partner, $py(dmgH)_2Co^{II}$-mediated β-H elimination, then rapid, effectively irreversible disproportionation of $py(dmgH)_2CoH$.[5] This general cross coupling mechanism is illustrated with α-ethoxyacrylonitrile in the following scheme.

$$R\text{-}Co^{III}(dmgH)_2py \underset{\longleftarrow}{\overset{\text{visible light}}{\longrightarrow}} R\cdot \quad + \quad \cdot Co^{II}(dmgH)_2py$$

$$R\cdot \quad \xrightarrow{} \quad \underset{OEt}{\overset{CN}{\diagup}} \quad \xrightarrow{} \quad R\underset{OEt}{\overset{CN}{\diagdown}}\cdot$$

$$\cdot Co^{II}(dmgH)_2py$$

$$H\text{-}Co^{III}(dmgH)_2py \quad \xrightarrow{} \quad R\underset{OEt}{\diagup}{}^{CN}$$

$$1/2\ H_2 \quad + \quad \cdot Co^{II}(dmgH)_2py \quad \xleftarrow{} \quad H\text{-}Co^{III}(dmgH)_2py$$

$$R\underset{OEt}{\overset{Co(dmgH)_2py}{-CN}} \quad \xrightarrow{\text{visible light}}$$

From a practical point of view, the reactions are easy to perform with equipment and techniques common in the synthetic organic laboratory. Primary and secondary alkyl cobaloximes can be prepared inexpensively (~$50/mole) from alkyl halides or sulfonate esters + in-situ-prepared $NaCo^I(dmgH)_2py$ "supernucleophile" via common bench-top inert atmosphere techniques. $RCo^{III}(dmgH)_2py$ are stable to brief exposure to air and room light. Thus, they can be handled as common synthetic organic intermediates in operations such as silica gel flash chromatography, rotary evaporation, and weighings on top-loading balances. Anaerobic visible light photolyses using 300 W incandescent light bulbs can be performed in common Pyrex glassware.

Acknowledgements: We acknowledge the financial support of the University of Oregon, an M. J. Murdoch Charitable Trust Grant of the Research Corporation, the Petroleum Research Fund administered by the American Chemical Society, and the Alfred P. Sloan Foundation. The General Electric QE-300 NMR spectrometer used in this work was purchased with funds provided by PHS RR 02336 and NSF CHE 8411177.

REFERENCES

1 Fellow of the Alfred P. Sloan Foundation 1987-1989
2 Branchaud BP, Meier MS, Malekzadeh MN (1987) J Org Chem 52:212
3 Branchaud BP, Meier MS, Choi YL (1988) Tetrahedron Lett 167
4 Branchaud BP, Meier MS Tetrahedron Lett in press
5 (a) Ng FTT, Rempel GL, Halpern J (1982) J Am Chem Soc 104:621
  (b) Gjerde HB, Espensen JH (1982) Organometallics 1:435
  (c) Halpern J, Ng FTT, Rempel GL (1979) J Am Chem Soc 101:7124

MAGNETIC EFFECTS IN THE
PHOTOCHEMISTRY OF KETONES

A.L. Buchachenko, E.N. Step,
V.F. Tarasov, and I.A. Shkrob

The Institute of Chemical Physics
of the Academy of Sciences of the USSR
117334 Moscow Kosygin street 4, USSR

Selection of Si isotopes (^{29}Si against ^{28}Si and ^{30}Si) induced by the magnetic isotope effect has been discovered under photolysis of silyl-containing ketones. The direct photolysis of $PhCH_2COSi(CH_3)_2Ph$ induces decomposition of the ketone via CH_2–CO bond in the singlet state, while sensitized photolysis leads to decomposition of the ketone via the same bond in the triplet state (with the probability of about 1/3) and to carbene insertion (with the probability of about 2/3). The radical mechanism of the decomposition is accompanied by enrichment of the initial ketone with magnetic isotope (under sensitized photolysis) and by its depletion (under direct photolysis). The photolysis of $PhCOSi(CH_3)_2Ph$ occurs via the mechanism of carbene insertion with no isotope selection.

The photolysis of optically active ketone $PhCOCH(CH_3)Ph$ is accompanied by racemization with the loss of optical activity being magnetically sensitive. The probabilities of radical pairs recombination and disproportionation as function of the external magnetic field and hyperfine interaction have been found.

H. Fischer, H. Heimgartner (Eds.)
Organic Free Radicals
© Springer-Verlag Berlin Heidelberg 1988

ESR STUDY Of REACTIVE INTERMEDIATES IN THE RADIOLYSIS AND PHOTOLYSIS OF AL ALKOXIDES

A.Faucitano,A.Buttafava,F.Martinotti
Dip. Di Chimica Generale V.le Taramelli 12   27100   PAVIA (Italy)

The ESR spectroscopy coupled with matrix isolation at low temperature and spin trapping is used to investigate the unstable species produced by electron capture, electron loss and homolytic bond rupture in gamma and uv irradiated AL alkoxides(1) .The results are aimed to contribute to the elucidation of the chemistry of the reactive intermediates of this class of organometallic compounds which is of interest in connection with their role as cathalyst . Al-isopropylate and Al-sec-Butylate from Fluka were used as received.Uv and gamma irradiation were performed at 77k under vacuum by using a 100 w high pressure Mercury lamp and a ^{60}Co gamma source.The esr spectra were recorded on a Varian E-109 spectrometer.2-methyl-2nitroso propane (MNP) was used as a spin trapping agent in isooctane and n-hexane solutions.

The uv irradiation at 77 k of neat Al-sec-butylate yields the ethyl radical pattern (2 a(H) = 22 G ,3 a(H)= 27 G)superimposed on a broad sextet of about 20 g splitting attributable to the hydrogen abstraction radical A (5 beta aproximately equivalent interacting protons).

$$CH_3(CH_3CH_2)\overset{\cdot}{C}OAl(OR)_2 \quad A$$

Correlated results are obtained from the 77 K photolysis of Al-isopropylate which gives an ESR spectrum showing the signal of methyl radicals (3 a(H) = 23 G) superimposed on a binomial septet of about 20 G attributable to the Hydrogen abstraction radical B

$$(CH_3)_2\overset{\cdot}{C}OAl(OR)_2 \quad B$$

The identification of these free radical products is consistent with the reaction sequence:

1- $(RO)_3Al \longrightarrow [RO^\cdot]^* + (RO)_2Al^\cdot$
2- $[RO^\cdot]^* \longrightarrow R^\cdot (CH_3^\cdot, CH_3CH_2^\cdot) + CH_3CHO$
3- $[RO^\cdot]^* + (RO)_3Al \longrightarrow ROH + A, B$

Reaction 3 may occurr intramolecularly because of the trimeric and tetrameric structure of the alkoxides.
The Esr signal of the species $(RO)_2Al^\cdot$ is not detected by esr in our experimental conditions ,however spin trapping with MNP applied to the photolysis of Al-sec-butylate in n-hexane solutions at 213 K yields the nitroxide C which is the spin adduct of the fragmentation radical $CH_3CH_2(CH_3)CH^\cdot$

$$CH_3CH_2(CH_3)CH-N-\overset{\cdot}{O}\cdots Al(OR)_3$$
$$|$$
$$t-But$$

a(N) =15.5 G   C
a(H)=15.5 G
a(Al)=1.6 G

This observation prompts the hypothesis that the intermediate $(RO)_2Al^\cdot$ radical decomposes according to the reaction :

4- $(RO)_2Al^\cdot \longrightarrow R^\cdot (CH_3CH_2\overset{\cdot}{C}HCH_3 , (CH_3)_2\overset{\cdot}{C}H) + ROAl=O$

H. Fischer, H. Heimgartner (Eds.)
Organic Free Radicals
© Springer-Verlag Berlin Heidelberg 1988

Upon warming to 293 K the spin adduct C decays irreversibly generating the nitroxide D which is the spin adduct of the hydrogen abstraction radical A :

$$CH_3CH_2(CH_3)C-OAl(OR)_2 \qquad a(N)=28. \text{ G}$$
$$\underset{|}{\text{ }} \qquad a(Al)=1.65 \text{ G}$$
$$t-But-N-O^\cdot$$

This conversion may tentatively be reckoned with the hydrogen abstraction reaction

$$\text{>}N-O^\cdot \ +Al(OR)_3 ----\blacktriangleright \ \text{>}N-OH \ + \ A$$
$$A \ + \ MNP ----\blacktriangleright \ \ D$$

The nitroxide D shows an unusually large N h.f. splitting which is diagnostic of a large departure from planarity of the nitrogen radical centre .This effect may be correlated with intramolecular complexation to a neighboring Al atom.

The esr spectra obtained fro Al-sec-Butylate after gamma irradiation at 77 K consist of a broad doublet of 70-80 G superimposed to a narrower singlet.On warming the doublet decays generating the pattern of the hydrogen abstraction radical A.The doublet has the characteristics expected from an oxygen centred cation radical (2,3) which may generate A by intramolecular hydrogen abstraction from the adjacent or nearby methyne groups.The spin trapping method applied to the radiolysis of Al-sec-Butylate yields the spin adduct C of the fragmentation radical A wich converts to the spin adduct D upon warming ;the h.f.parameters of the nitroxides are identical to those observed in the photolysis experiments.

The gamma irradiation of neat Al-isopropylate at 77 K followed by annealing to R.T. yields the hydrogen abstraction radical B,showing a well resolved coupling to one Al atom , as the major radiolytic product.

$$(CH_3)_2\overset{\cdot}{C}O-Al(OR)_3 \qquad 6 \ a(H)=19.1 \text{ G} \qquad a(Al)=4.7 \text{ G}$$

Spin trapping with MNP yields the spin adduct of alkoxy radicals

$$(CH_3)_2CHO-N-O^\cdot \qquad a(N)=28. \text{ G}$$
$$\underset{t-But}{|}$$

togheter with the adduct D of the hydrogen abstraction radical .

The Al-O ,C-O and C-H bond scissions taking place under gamma radiations may be rationalized through a mechanism initiated by electron capture electron loss events as suggested in the following scheme:

$$5- \quad Al(OR)_3 ------\blacktriangleright Al(OR)_3^{+\cdot} \ , \quad Al(OR)_3^{-\cdot}$$
$$6- \quad Al(OR)_3^{+} \ \llcorner-----\blacktriangleright \ ^\cdot Al(OR)_2 +RO^{-}$$
$$7- \quad R-O^{+}-Al(OR)_2 \ ---\blacktriangleright \ A,B \ + \ ^\cdot RO^{+}(H)Al(OR)_2$$
$$8- \quad (RO)_3Al^{+\cdot} \ + \ e^{-} \ -----\blacktriangleright \ [(RO)_3Al]^{\cdot} \ ---\blacktriangleright \ ^\cdot Al(OR)_2 \ + \ RO^{\cdot}$$
$$9- \quad ^\cdot Al(OR)_2 \ ----\blacktriangleright (RO)Al=O \ +R^{\cdot}$$
$$10- \quad RO^{\cdot} \ + \ Al(OR)_3 \ ---\blacktriangleright \ \ ROH \ + \ A,B$$

REFERENCES

1 Faucitano A.,Buttafava A.,Martinotti F.,Bortolus P.,Comincioli,V. Proceedings Conv. Ital. Sci. Macromol. 7th (1985),111
2 Wang,J.T.,Williams F. J.Amer.Chem.Soc. 103,6994,(1981)
3 Symons M.C.R.,Wren B.W. Chem.Comm. 15,817,(1982)

RATE CONSTANTS FOR ELECTRON TRANSFER AND
RADICAL CATION REACTIONS OF METHYLATED NAPHTHALENES

W.A. Rodin, S. Frye, and D.M. Camaioni

Battelle, Pacific Northwest Laboratories
P.O. Box 999, Richland, WA 99352

Abstract: Since kinetic measurements of base-promoted arene oxidations by tris(phenanthroline)iron(III) provide a ready method for determining rate constants for both electron transfer (ET) and follow-up reactions of arene radical cation intermediates, [1-3] we are using this approach to study the effects of structure on the reactions of radical cations, especially the competition between side chain fragmentation and nucleophilic addition reactions. In this paper, we report our findings for methylated naphthalenes and make comparisons with data for methylated benzenes that were reported by Kochi, Amatore and Schlesener [1].

$$Fe(III) + ArCH_3 \underset{k_{-1}}{\overset{k_1}{\rightleftarrows}} Fe(II) + ArCH_3^{+\bullet} \overset{k_2}{\underset{Pyr}{\longrightarrow}} \begin{array}{l} \longrightarrow ArCH_2\bullet \ + \ Pyr\text{-}H^+ \\[1em] \longrightarrow Pyr^+\underset{H}{\overset{\bullet}{>}}ArCH_3 \end{array}$$

$$Fe(III) \ + \ R\bullet \ \overset{fast}{\longrightarrow} \ Fe(II) \ + \ R^+ \ \overset{Pyr}{\longrightarrow} \ R\text{-}Pyr^+$$

Experimentally, rate constants for ET (k_1) and for radical cation reactions relative to back electron transfer (k_2/k_{-1}) are obtained by monitoring the appearance of Fe(II) spectrophotometrically at 510 nm ($\epsilon = 11,400$) and fitting the time-dependent absorbance data to the integrated rate equation using multiple linear regression analysis [2]. If oxidation potentials for arenes are known, then values for k^{-1} and k^2 can be determined from the equilibrium constants for ET, that are calculated from the Nerst equation, $K = k_1/k_{-1} = (E°_{Ar}-E°_{Fe})\mathcal{F}/RT$, where $E°_{Fe} = 1.09$ V (SCE) [2] and $\mathcal{F}$ is Faraday's constant (23.06 kcal/V).

$$-\frac{k_{-1}\epsilon A^\infty + k_2[B]}{2k_1k_2[Ar][B]} \ln \frac{A - A°}{A^\infty - A°} - \frac{k_{-1}\epsilon(A^\infty - A°)}{2k_1k_2[Ar][B]}\left(1 - \frac{A - A°}{A^\infty - A°}\right) = t$$

Table 1 lists results for mono- and di-substituted methylnapthalenes obtained using pyridine and lutidine bases to promote the oxidations. Cyclic voltammetry peak potentials measured by Yoshida and Nagase [4] are included in the table to establish the relative thermodynamic driving forces for oxidation of the arenes (the smaller the potential, the greater the driving force). The ET rates increase substantially on going from 2MNP to 1,8DMN and the rates parallel the driving force for oxidation.

H. Fischer, H. Heimgartner (Eds.)
Organic Free Radicals
© Springer-Verlag Berlin Heidelberg 1988

The relative rates, k_{-1}/k_2, increase, too, with the driving force for oxidation, and suggest that reactivity of the radical cations toward base decreases as the stability of the radical cations increases. Interestingly, the radical cations react from 2-20 times _faster_ with pyridine than with the _stronger_ base, lutidine, which indicates that pyridine reacts with these radical cations mainly by nucleophilic addition, while lutidine, a sterically hindered nucleophile, reacts by proton transfer. In the last two columns of Table 1, relative reactivities of the arene cation radicals towards lutidine and pyridine are shown as determined using the Nernst equation and k_1 values to estimate k_{-2}. The data show that the reactivity of the radical cations towards pyridine is less sensitive to radical cation stability than is the reactivity of lutidine. This behavior, too, is more consistent with pyridine reacting by nucleophilic addition rather than proton transfer with these arenes.

Table 1. Rate Data for oxidation of Methylated Naphthalenes by Iron(III).[a]

| Arene | E_p, V[b] | k_1, $M^{-1}s^{-1}$ | k_{-1}/k_2 | | $\dfrac{k_2 Pyr}{k_2 Lut}$ | $k_2 rel$ | |
			Lut	Pyr		Lut	Pyr
2MNP	1.58	0.70 (0.10)	190 (63)	86 (45)	2.2	1.0	1.0
1MNP	1.54	1.5 (0.4)	880 (340)	57 (25)	15	0.097	0.68
2,3DMN	1.49	6.2 (1.0)	440 (160)	60 (34)	7.3	0.12	0.38
2,6DMN	1.47	18 (2)	1300 (330)	70 (19)	19	0.052	0.44
1,5DMN	1.46	60 (7)	6300 (1500)	570 (140)	11	0.24	0.12
1,4DMN	1.44	160 (40)	6700 (2500)			0.28	
1,8DMN	1.41	470 (90)	13000 (3200)	840 (140)	15	0.013	0.092

[a]Tris(phenanthroline)iron(III) in acetonitrile; MNP: methylnaphthalene; DMN: dimethylnaphthalene. Numbers in parentheses indicate one standard deviation.
[b]Cyclic voltammetry peak potentials vs. SCE in methanol, ref 4.

Absolute rates for k_{-1} are estimated to be in the range, $2-6 \times 10^7$ $M^{-1}s^{-1}$, which allows for the radical cations to exhibit reaction rates of 10^3-10^5 $M^{-1}s^{-1}$ with lutidine and 10^5-10^6 $M^{-1}s^{-1}$ with pyridine. For comparison, hexa-, penta- and tetramethylbenzenes studied by Kochi and coworkers [1, 2], have $E° = 1.7-1.5$ V (SCE) and exhibit values for k_1 ranging 0.006-14 $M^{-1}s^{-1}$, values for k_{-1} ranging 10^8-10^9 $M^{-1}s^{-1}$ and values for k_2 ranging 10^5-10^6 $M^{-1}s^{-1}$ with pyridine and 10^6-10^7 $M^{-1}s^{-1}$ with lutidine. Thus, data presented here extend the range of reactivities that have been determined for arene radical cations by this approach and allow for determining the effects of aromatic structure on ET and radical cation reaction of arenes.

1   Schlesener CJ, Amatore C, Kochi JK (1986) J Phys Chem 90:3747
2   Schlesener, CJ, Kochi JK, (1984) J Org Chem 49:3142
3   Schlesener, CJ, Amatore C, Kochi J K (1984) J Am Chem Soc 106:3567,7472
4   Yoshida k, Nagase S (1979) J Am Chem Soc 101:4268

CYCLOPENTANE RING FORMATION IN THE CYCLOADDITION REACTIONS OF 3-BUTENYL RADICAL ON THE ELECTRON-DEFICINET OLEFINS

Živorad Čeković[*] and Radomir Saičić

Department of Chemistry, Faculty of Sciences, University of Belgrade
Studentski trg 16, 11001 Belgrade, Yugoslavia

Abstract: By decomposition of thiohydroxamic esters of 4-pentenoic acid in the presence of symmetrical electron-deficient olefinic compounds, intermolecular and intramolecular radical additions takes place and polysubstituted cyclopentane derivatives were obtained.

Intramolecular addition of alkenyl radicals takes place independently of the radical precursors, but depending of the reagents and reaction conditions different functionality of the cyclopentane derivatives were observed[1-4].

We investigated intermolecular addition of 3-butenyl radical, on the symmetrical electron-deficient olefinic compounds 1a-c, that were used as a radical acceptors. 3-Butenyl radical was generated by decomposition of thiohydroxamic ester of 4-pentenoic acid 2 under thermal conditions[5]. Intermolecular addition was followed by intramolecular addition reaction and polysubstituted cyclopentane derivatives 3, possessing two functional groups and functionalized three carbon atom side chain, have been obtained in yields of 55-63% (Scheme 1.).

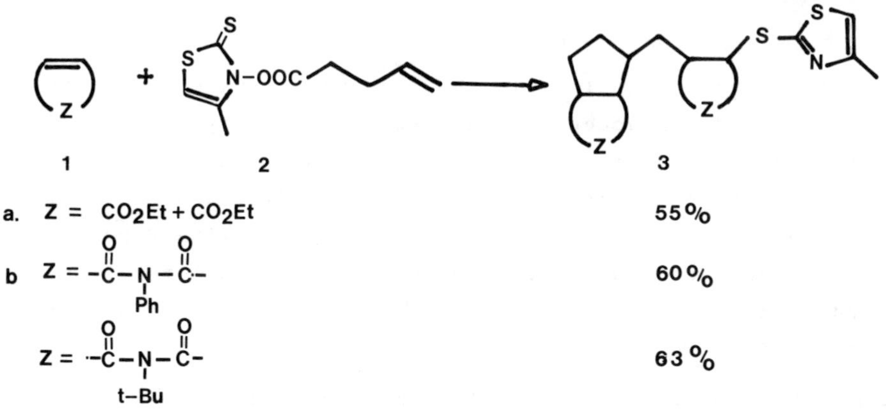

a. Z = CO$_2$Et + CO$_2$Et		55%
b. Z = -C(O)-N(Ph)-C(O)-		60%
Z = -C(O)-N(t-Bu)-C(O)-		63%

Scheme 1.

Reactions were carried out by slowly addition of one equivalent of thiohydroxamic ester of 4-pentenoic acid 2 to the boiled solution of 3 equivalents of olefinic radical acceptors 1 in toluene. The reac-

H. Fischer, H. Heimgartner (Eds.)
Organic Free Radicals
© Springer-Verlag Berlin Heidelberg 1988

tion mixture was refluxed during additional 30 minutes. Lower yields
of cycloaddition products were obtained when reactants were mixed
and refluxed. The reaction products were separated by column chromatography by using benzene : ethyl acetate (8 : 2) as an eluent. The
reaction products were characterized by ir, nmr and mass spectra.

This free radical chain reaction was initiated by thermaly induced homopolar cleavage og N-O bond of thiohydroxamic ester 2. By decarboxylation of 4-pentenoyloxy radical 3-butenyl radical was generated. This alkenyl radical undergoes to the intermolecular addition
on the olefinic bond of maleic acid derivatives 1 (Scheme 2). 5-Hex-
enyl type radical 4, arised from the intermolecular addition reaction,
possessing an appropriately oriented olefinic bond, undergoes to the
5-exo-cyclization thus closing the cyclopentane ring and generating

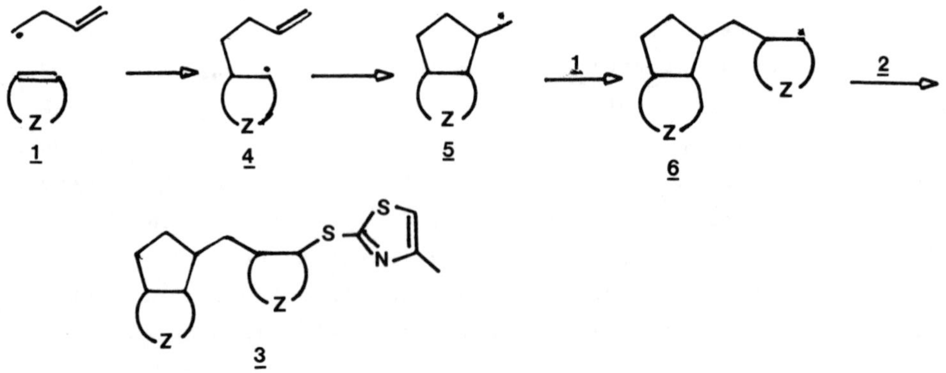

Scheme 2.

the cyclopentylmethyl radical 5. In the presence of excess of olefinic
compounds 1 as a radical acceptors the cyclopentylmethyl radical 5
undergoes to the new addition reaction to olefinic compound 1 thus
forming a radical 6. Termination step involves addition of radical 6
on the C S bond of the starting thiohydroxamic ester with formation
of polysubstituted cyclopentane derivatives 3 as a final reaction
products and a new 4-pentenoyloxy radical arised which continue this
free radical chain reaction.

1) M. Ramaiah, Tetrahedron, 43, 3541 (1987) and references therein
2) G. Stork and P. M. Sher, J. Am. Chem. Soc., 108, 303 (1986)
3) Ž. Čeković and R. Saičić, Tetrahedron Letters, 27, 5893 (1986).
4) Ž. Čeković and D. Ilijev, Tetrahedron Letters, 29, 1441 (1988)
5) D. H. R. Barton and G. Kretzschmar, Tetrahedron Letters, 5889 (1983).

α-HALOSULFONES REDUCTION MECHANISM :
APPLICATION OF HIGHLY EFFICIENT RADICAL CLOCKS

B. Vacher, A. Samat and M. Chanon

UA 126 CNRS. Faculté des Sciences St Jérôme.B 561.
13397 Marseille Cédex 13. France

The synthesis of highly efficient radical clocks of the 5-*(exo)*-substituted-5-*(endo)*-isopropyl sulfonyl type was recently developped[1]. The rates of the intramolecular radical cyclization (reaction (1)) are presently among the fastest (10^8 and 10^9 s^{-1} respectively for R=CN and R=C$_6$H$_5$).

The compound **1** (X = Cl or Br) was used to study the mechanism of the reduction of α–halosulfones by Zn or $SO_3^=$ (reaction(2)). Indeed for these reactions in protic solvent both ionic[2] and electron transfer mechanisms[3] have been proposed.

When methanol is the solvent, traces of tricyclic compound **3** (1 to 3%) appear as products of the reaction. The amount of **3** is greatest (untill to 18%) when α-bromosulfone is reduced in HMPA as solvent.

The presence of traces of the tricyclic compound **3** during the experiments performed show the participation of paramagnetic species. Then the reaction may process by a double single electron transfer mechanism as postulated by C.Y. Meyers[3] et al. : the α–sulfonyl carbanion is formed in a first step, followed by two fast electron transfers occuring in a solvent cage. Do the ionic and electron transfer pathways coexist?

1 Vacher B, Samat A, Chanon M (1985) Tetrahedron Lett 26:5129; Vacher B, Samat A, Chanon M....(1988) in press; Vacher B (1987) Sciences Thesis, Marseille

2 Bordwell FG, Doomes E (1974) J Org Chem 39: 2298

3 Hua DH (1979) Ph. D. Thesis, S.I.U. Carbondale, p 94

H. Fischer, H. Heimgartner (Eds.)
Organic Free Radicals
© Springer-Verlag Berlin Heidelberg 1988

TRIS(TRIMETHYLSILYL)SILANE.
A NEW REDUCING AGENT

C. Chatgilialoglu[¶] and D. Griller[†]

[¶]I.Co.C.E.A., C.N.R., Ozzano Emilia (Bologna), Italy and
[†]Division of Chemistry, National Research Council, Canada

Trialkylsilanes are poor reducing agents in a free radical chain reactions. For example, although trialkylsilyl radicals are very reactive in halogen atom abstraction(1) their corresponding silanes are rather poor H-atom donors towards alkyl radicals and therefore not do support chain reactions except at elevated temperatures(2). Recently we found that the silicon-hydrogen bonds can be dramatically weakened by multiple substitution of Me_3Si groups at the Si-H function. In fact the silicon-hydrogen bond in $(Me_3Si)_3SiH$ is 79 kcal mol^{-1} and is 11 kcal mol^{-1} less than that of $Et_3SiH(3)$. This result suggests that tris(trimethylsilyl)silane might be a good hydrogen donor and that the compound would be capable of sustained radical chain reduction of organic halides. This expectation turned out to be correct. The reduction of alkyl halides by tris(trimethylsilyl)silane can be achieved under a variety of conditions in excellent yields(4). This reduction reaction involves a two-step free radical chain process, viz.,

$$R^{\cdot} \;+\; (Me_3Si)_3SiH \;\longrightarrow\; RH \;+\; (Me_3Si)_3Si^{\cdot} \qquad (1)$$

$$(Me_3Si)_3Si^{\cdot} \;+\; RX \;\longrightarrow\; (Me_3Si)_3SiX \;+\; R^{\cdot} \qquad (2)$$

as indicated by the fact that the reaction is catalysed by thermal or photochemical sources of free radicals and retarded by common inibitors. The reduction were very efficient for alkyl iodides, bromides and chlorides with the following reactivity order: RI > RBr > RCl.

A method for catalytic dehalogenation via tris(trimethylsilyl)-silane was also developed: the organic halide is treated with excess of sodium borohydride and a catalytic amount of $(Me_3Si)_3SiH$ or its corresponding halide, the reaction being initiated photochemically(5).

H. Fischer, H. Heimgartner (Eds.)
Organic Free Radicals
© Springer-Verlag Berlin Heidelberg 1988

The catalytic cycle is given in eq. 3 and 4.

$$(Me_3Si)_3SiX + NaBH_4 \longrightarrow (Me_3Si)_3SiH + NaX + BH_3 \quad (3)$$

$$(Me_3Si)_3SiH + RX \longrightarrow (Me_3Si)_3SiX + RH \quad (4)$$

Summarizing, tris(trimethylsilyl)silane appear to offer an attractive alternative to tributyltin hydride in radical chain reactions.

1 Chatgilialoglu C, Ingold KU, Scaiano JC (1982) J Am Chem Soc 104:5123

2 Lusztyk J, Maillard B, Ingold KU (1986) J Org Chem 51:2457

3 Kanabus-Kaminska JM, Hawari JA, Griller D, Chatgilialoglu C (1987) J Am Chem Soc 109:5267

4 Chatgilialoglu C, Griller D, Lesage M (1988) J Org Chem 53:0000

5 Chatgilialoglu C, Griller D, Lesage M (1987) Italian Patent:48150A87

AUTOXIDATION OF $(Me_3Si)_3SiH$

C. Chatgilialoglu and G. Seconi

I.Co.C.E.A., C.N.R., Ozzano Emilia (Bologna) Italy

The oxidation of trialkyl- and triarylsilanes induced by free radical initiators gave silanols as the major product, and no species corresponding to a hydroperoxide was isolated, in contrast with the autoxidation of the carbon analogues. Possible reaction schemes for such behaviour have been proposed in the literature.

In this communication we report a study on the autoxidation of tris(trimethylsilyl)silane. The atmospheric oxidation of $(Me_3Si)_3SiH$ is a relatively fast process, viz.,

$$(Me_3Si)_3SiH + O_2 \longrightarrow Me_3SiSi(H)(OSiMe_3)_2 \tag{1}$$

Up to 50% conversion of $(Me_3Si)_3SiH$ the yields are essentially quantitative after which they decrease. Experimental evidence suggests that the following unexpected "intramolecular" free radical chain process is responsible for such behaviour:

$$(Me_3Si)_3Si^\bullet + O_2 \longrightarrow (Me_3Si)_3SiOO^\bullet \tag{2}$$

$$(Me_3Si)_3SiOO^\bullet \xrightarrow{\text{1,3-shift}} (Me_3Si)_2\overset{\bullet}{Si}OOSiMe_3 \tag{3}$$

$$(Me_3Si)_2\overset{\bullet}{Si}OOSiMe_3 \xrightarrow{\underset{i}{S}H} (Me_3Si)_2\overset{\bullet}{Si}(O)OSiMe_3 \tag{4}$$

$$(Me_3Si)_2\overset{\bullet}{Si}(O)OSiMe_3 \xrightarrow{\text{1,2-shift}} Me_3Si\overset{\bullet}{Si}(OSiMe_3)_2 \tag{5}$$

$$Me_3Si\overset{\bullet}{Si}(OSiMe_3)_2 + (Me_3Si)_3SiH \longrightarrow \tag{6}$$

$$\longrightarrow Me_3SiSi(H)(OSiMe_3)_2 + (Me_3Si)_3Si^\bullet$$

To our knowledge the unimolecular steps 3, 4 and 5 are unknown reactions. We believe that the strength of the silicon-oxygen bond is a potent driving force in these novel rearrangements.

H. Fischer, H. Heimgartner (Eds.)
Organic Free Radicals
© Springer-Verlag Berlin Heidelberg 1988

Ti(III) SALT MEDIATED RADICAL ADDITION TO CARBONYL CARBON.

SYNTHESIS OF α,β-DIHYDROXYKETONES.

Angelo Clerici and Ombretta Porta

Dipartimento di Chimica del Politecnico,
Pza L. da Vinci, 32 - 20133 Milano, Italy

Abstract: Carbon centered radicals 2, generated by reduction of α,β-dicarbonyl compounds 1 by Ti(III) salt, add to the carbonyl carbon of aldehydes 3 to afford α,β-dihydroxyketones 5 in reasonable to high yields. Good stereoselectivity, depending on the nature of the reactants, is achieved.

Based on the knowledge of carbon-to-carbon bond-forming reactions we have previously explored[1], the one-pot synthesis of α,β-dihydroxyketones 5 is realized by allowing to react 1, 3, and aqueous $TiCl_3$ in the ratio 1:1:2 per 1 h at 0 °C in acetone or THF as solvent.

$R= R_1=$ Ph, Furyl;　$R=$ Ph, $R_1=$ H;　$R=$ Ph, $R_1=$ CH_3.　(65-100% yield)

$R_2=$ CH_3, C_2H_5, $(CH_3)_2CH$, Ph, o-OH-Ph, o-CH_3Ph, o-OCH_3Ph, o-COOH-Ph, $PhCH_2$, $PhCH_2CH_2$, Furyl.

α,β-Dihydroxyketones 5 are obtained as a mixture of two isomers, one of which strongly predominate when R_2 and/or R_1 are sterically demanding, to the point that only one stereomer is detectable when $R_2=R_1=$ Ph or benzaldehyde is ortho-substituted with a group capable of chelation with the Ti(III) ion (e.g.; OH, COOH).

Since radical addition to the carbonyl carbon is a reversible process, the reduction of the intermediate alkoxy radicals 4 by Ti(III) ion accounts for the driwing force of the reaction.

1. A. Clerici and O. Porta, J. Org. Chem., 47, 2852 (1982); A. Clerici and O. Porta, Ibid., 48, 1690 (1983); A. Clerici and O. Porta, Tetrahedron, 38, 1293 (1982); A. Clerici and O. Porta, Ibid., 46, 561 (1986); A. Clerici and O. Porta, J. Org. Chem., 52, 5099 (1987).

H. Fischer, H. Heimgartner (Eds.)
Organic Free Radicals
© Springer-Verlag Berlin Heidelberg 1988

Polarized EPR Spectra of Unstable Biradicals

Gerhard L. Closs
Department of Chemistry
The University of Chicago
Chicago, Illinois

Abstract: Laser flash photolysis of α,α'-methylated cyclic ketones produces acyl-alkyl biradicals via Norrish type I cleavage. Subsequent thermo-activated decarbonylation yields dialkyl biradicals. EPR signals are obtained by direct detection with a boxcar averager. The spectra show strong polarization of a type different from conventional CIDEP spectra. The spectra have been analyzed and simulated by computing the eigenvalues and eigenvectors of the appropriate biradical spin Hamiltonian containing Zeeman, exchange and hyperfine interactions. The kinetics model is based on the assumption that product formation is proportional to the singlet character of the individual hyperfine states. It also incorporates spin relaxation. Both acyl-alkyl and alkyl-alkyl types biradicals with chain lengths from 7 through 18 carbon atoms have been observed and their spectra simulated. The magnitude of the exchange interaction is analyzed with the help of Monte Carlo calculations on a rotational isomeric state model. Attempts to fit the chain length dependence of the exchange interaction with a through-space model were unsuccessful. A satisfactory model does include through-bond interaction for the shorter chains.

ATOM-TRANSFER ADDITION, CYCLIZATION, AND ANNULATION

Dennis P. Curran,
Department of Chemistry, University of Pittsburgh,
Pittsburgh, PA 15260, USA

Sequential free radical reactions have been used with considerable success to rapidly assemble functionalized organic molecules. Often, such reactions are carried out with a trialkyltin hydride reagent. However, the use of tin hydride is not always desirable (or possible), particularly when one or more slow steps are involved in a given sequence. The unique capability of iodine atom transfer reactions to control free radical sequences will be discussed. In this "atom transfer" method, the starting iodide substitutes for the tin hydride in the key chain transfer step. With proper design, intermediate radicals are permitted maximum lifetimes (to undergo desired reactions) while final radicals are permitted very short lifetimes. The development of addition, cyclization, and annulation reactions controlled by iodine atom transfer will be described. Several representative examples which focus on the chemistry of iodomalonates are shown below.

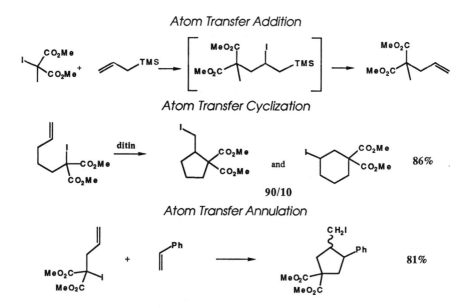

H. Fischer, H. Heimgartner (Eds.)
Organic Free Radicals
© Springer-Verlag Berlin Heidelberg 1988

Representative Publications

D. P. Curran, The Design and Application of Free Radical Chain Reactions in Organic Synthesis, *Synthesis* (review, in press).

D. P. Curran and M.-H. Chen, Atom Transfer Cycloaddition. A Facile Preparation of Functionalized (Methylene)cyclopentanes. *J. Am. Chem. Soc.* **1987**, *109*, 6558.

D. P. Curran and T.-C. Chang, Atom Transfer Cyclization Reactions of α-Iodo Carbonyls, *Tetrahedron Lett.* **1987**, *28*, 2477.

D. P. Curran and D. Kim, Atom Transfer Cyclization of Simple Hexenyl Iodides. A Caution on the Use of Alkenyl Iodides as Probes for the Detection of Free Radical Intermediates. *Tetrahedron Lett.* **1986**, *27*, 5821.

D. P. Curran, M.-H. Chen, and D. Kim, Atom Transfer Cyclization. A Novel Isomerization of Hexynyl Iodides to Cyclic Vinyl Iodides. *J. Am. Chem. Soc.* **1986**, *108*, 2489.

MERCURIC OXIDE - IODINE OXIDATION OF
STEROIDAL HOMOALLYLIC ALCOHOLS

M. Lj. Mihailović, M. Dabović, M. Bjelaković and Lj. Lorenc

Department of Chemistry, Faculty of Science, University of Belgrade, Studentski trg 16, P.O. Box 550, YU-11001 Belgrade, and Institute of Chemistry, Technology and Metallurgy, Belgrade, Yugoslavia

Our previous study has shown that the oxidation of androst-4-ene-$3\beta,9\alpha,17\beta$-triol 3,17-diacetate (1) with mercuric oxide - iodine results predominantly in α-epoxidation of the olefinic double bond to produce, as the only detectable reaction product, the $4\alpha,5\alpha$-epoxy derivative (2) (in about 60% yield) (Scheme).

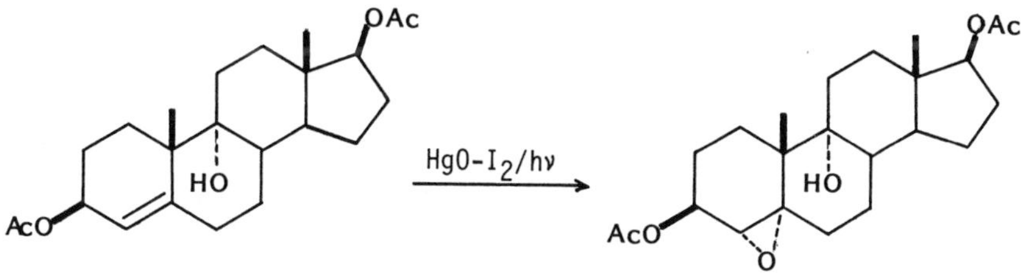

$$\underline{1} \qquad\qquad\qquad \underline{2}$$

Scheme

On the other hand, when the HgO - I_2 oxidation was performed with the corresponding 9α-desoxy analogue, i.e. androst-4-ene-$3\beta,17\beta$-diol diacetate, it gave a mixture of $4\alpha,5\alpha$- and $4\beta,5\beta$-epoxides in not over 17% yield, indicating that the 9α-hydroxy group in (1) plays an important role in oxygen addition, affecting both the stereochemistry and the efficiency of epoxidation of the olefinic Δ^4-double bond in compound (1).

Therefore, in order to get more information concerning the mechanism of this unusual epoxidation process with the HgO-I_2 combination, in this work similar oxidations have been performed starting from structurally different homoallylic steroidal alcohols, i.e. 5α-cholest-7-ene-$3\beta,5$-diol 3-acetate and cholest-5-ene-$1\alpha,3\beta$-diol 3-acetate. The results obtained will be presented and discussed.

H. Fischer, H. Heimgartner (Eds.)
Organic Free Radicals
© Springer-Verlag Berlin Heidelberg 1988

THE MECHANISMS OF THE REARRANGEMENTS
OF ALLYLIC HYDROPEROXIDES

A.L.J. Beckwith[a], A.G. Davies[b] and I.G.E. Davison[b].

[a]Research School of Chemistry, Australian National University, Canberra, A.C.T 2601, Australia

[b]Chemistry Department, University College London, 20 Gordon Street, London WC1H OAJ, U.K.

Cholesterol-5α-hydroperoxide (2) obtained by the singlet oxygenation of cholesterol (1), rearranges in chloroform to the 7α-hydroperoxide[1] (3) which then epimerises more slowly into the 7β-hydroperoxide[2] (4).

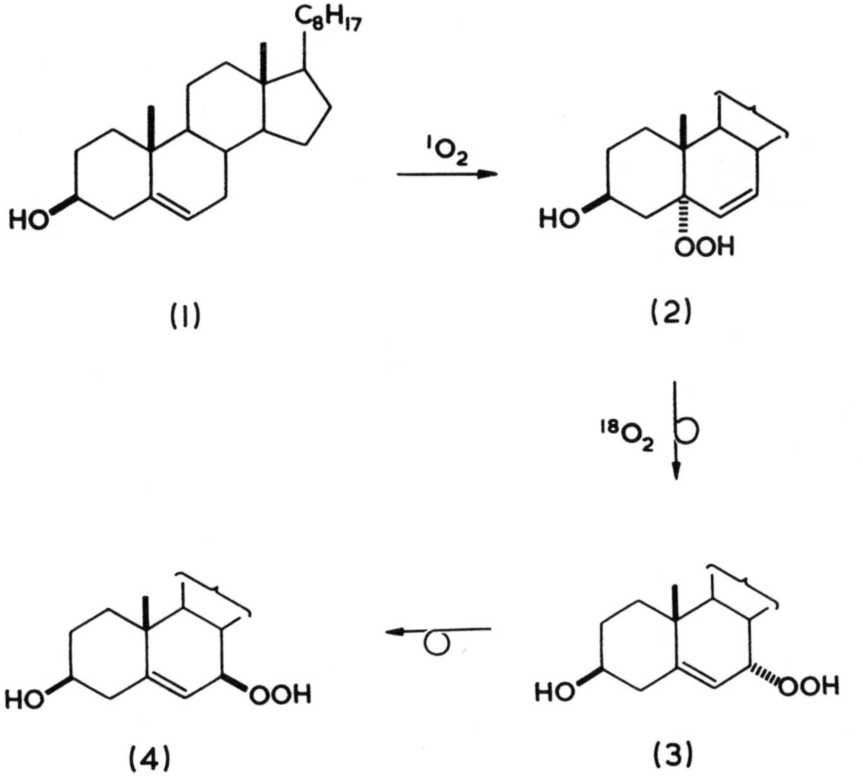

H. Fischer, H. Heimgartner (Eds.)
Organic Free Radicals
© Springer-Verlag Berlin Heidelberg 1988

If (2) is allowed to rearrange under an atmosphere of $^{18}O_2$, (3) is found to be isotopically normal, but (4) is heavily labelled with the ^{18}O isotope. It is suggested that the reaction (2) ⟶ (3) involves a concerted (sigmatropic) mechanism, but that the epimerisation (3) ⟶ (4) is a dissociative process[3].

The results of a kinetic study of the rearrangement (2) ⟶ (3) will be reported.

REFERENCES

(1) G.O. Schenck, O.A. Neumüller, and W. Enfield, Annalen, 1958, **618**, 202; B. Lythgoe and S. Trippett, J. Chem. Soc., 1959, 471.

(2) J.I. Teng, M.J. Kulig, L.L. Smith, G. Kan, and J.E. Van Lier, J. Org. Chem., 1973, **38**, 119.

(3) A.L.J. Beckwith, A.G. Davies, I.G.E. Davison, A.Maccoll, and M.H. Mruzek, J. Chem. Soc., Chem. Commun., 1988, 476.

STEREOSELECTIVE C-C BOND FORMATION IN CARBOHYDRATES BY RADICAL CYCLIZATION REACTIONS.

Alain De Mesmaeker, Beat Ernst

Ciba-Geigy Ltd., Central Research Laboratories, 4002 Basel, Switzerland

Our strategy for the stereoselective C-C bond formation in carbohydrates is based on intramolecular radical cyclization reactions and offers three major advantages in comparison with the intermolecular processes: a) a more efficient C-C bond formation b) a higher stereoselectivity due to the exclusive formation of a cis ring junction c) only a stoechiometric amount of the radical acceptor (C=C, C≡C) is necessary (in contrast with the large excess normaly used in the intermolecular processes).

The two approaches that we investigated are depicted below. In the first one, the radical acceptor (C=C, C≡C) is introduced selectively on a free hydroxylic function of the carbohydrate moiety having a radical precursor Y in the β position. An efficient radical cyclization occurs in a 5-exo mode yielding only the cis ring junction. Finally, the -X-O- bond is cleaved.

In the second approach, the radical acceptor is an endocyclic C=C double bond of the carbohydrate derivative.

The net result of both methods is the formation of a new C-C bond in the β position and cis to the initial hydroxyl group.

Examples of stereoselective C-C bond formations in C_1 and C_2 are given.

H. Fischer, H. Heimgartner (Eds.)
Organic Free Radicals
© Springer-Verlag Berlin Heidelberg 1988

A NOVEL REACTION OF TETRACYANOETHYLENE
WITH N-ARYL-ISOINDOLINES [1]

D. Döpp[a], A. A. Hassan[a], A. M. Nour-el-Din[b] and A. E. Mourad[b]

[a] Fachgebiet Organische Chemie, Universität Duisburg, D-4100 Duisburg
[b] Chemistry Dept., The University, El Minia, Arab. Republic of Egypt

Abstract: Tetracyanoethylene reacts with N-arylisoindolines 1 in aerated benzene under α-H-atom abstraction and formation of α-dicyanomethylene derivatives and α-oxygenated by-products.

Tetracyanoethylene (TCNE) normally reacts with an N,N-dialkylaniline (with free p-position) to products of p-tricyanovinylation; indole would undergo this reaction at C-3, primary aromatic amines on nitrogen [2]. Compounds with active methylene groups perform a Micheal-type addition, followed by elimination of malononitrile and formation of dicyanomethylene derivatives [2].

We found that TCNE and N-arylisoindolines (1, Ar = C_6H_5, $R-C_6H_4$ with R = o-, m-, p-OCH_3 or CH_3 or Cl and p-$COCH_3$) in benzene solution form intermediate charge transfer complexes which, in the presence of air oxygen at room temperature, give way to formation of products 8 - 11. When doubled molar amounts of TCNE are applied, the main products 8 are formed in up to 40% yield based on the amount of isoindoline 1 used. This is quite acceptable for a multistep conversion. The Z-structure of 8 (Ar = 3-CH_3-C_6H_4-) has been unambiguously proven by an x-ray structural analysis [3], a 37.7° twist around the central double bond seems to be noteworthy. When Ar = m- or o-substituted phenyl, some products of "normal" tricyanovinylation [1] may also be observed.

The necessarily complex reaction is tentatively and in abbreviated form outlined in the scheme. The formation of a symmetric product as 8 in the main pathway strongly suggests that none of the typical reactions of TCNE mentioned above and elsewhere [2,4] can serve as a precedent. Practically, 8, 9 and 11 can be looked at as products of α-oxidation of 1, and the formation of dicyanomethylene derivatives as 8 and 11 would - as a net result - formally constitute an "Umpolung" of the isoindoline α-carbons, which profit from the additional benzylic activations. There are ambiguities and alternatives, though, e. g. for the formation of 11 (directly from 7 or via 8) 5, which need clarification. Details will be published in forthcoming papers [3,5].

H. Fischer, H. Heimgartner (Eds.)
Organic Free Radicals
© Springer-Verlag Berlin Heidelberg 1988

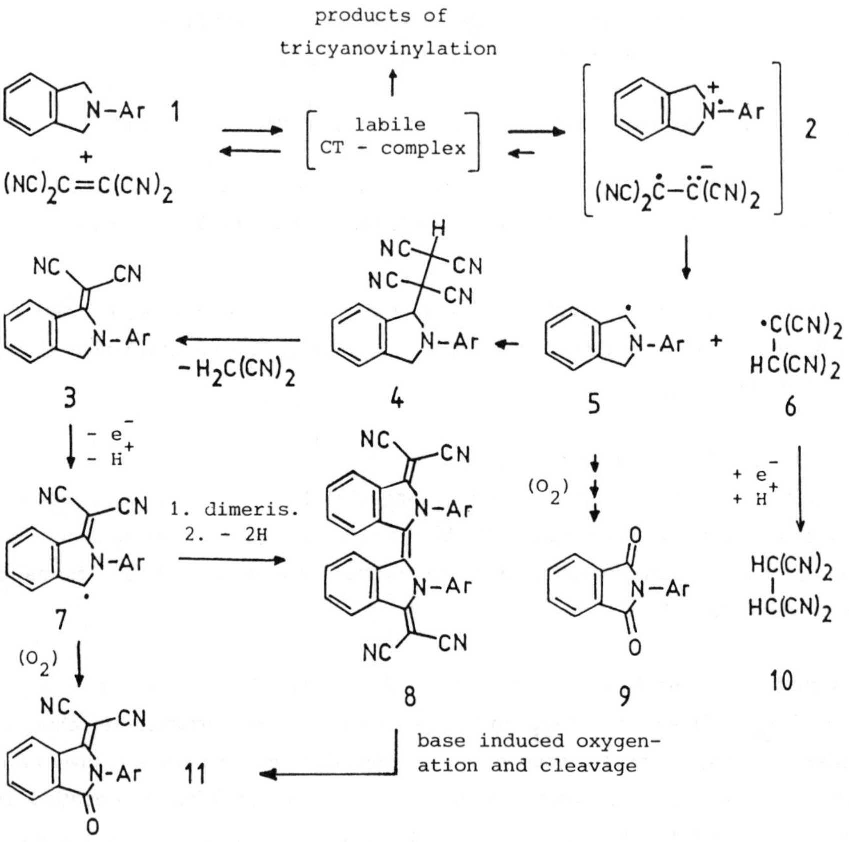

The electron acceptor properties of the deep blue isoindigoid compounds **8** are under investigation.

Acknowledgement: A. A. Hassan is endebted to the Egyptian government for a research fellowship under the channel system. Support by Fonds der Chemischen Industrie ist gratefully acknowledged.

References:

1   Taken in part from: Hassan A A (1987) PhD thesis, Univ. El Minia
2   Fatiadi A J (1986) Synthesis 1986: 249 and refs cited therein
3   Döpp D, Hassan A A, Krüger C, Angermund K (1987), in preparation
4   Fatiadi A J (1987), Synthesis 1987: 749 and refs cited therein
5   Döpp D, Hassan A A, Alberti A, Greci L, in preparation

AN ELECTRON TRANSFER INITIATED ISOMERIZATION
AND SOLVENT ADDITION VIA RADICAL CATION CHAIN PROCESS [1]

D. Döpp and H. Kretz

Fachgebiet Organische Chemie, Universität Duisburg,
Postfach 10 16 29, D-4100 Duisburg 1, Federal Republic of Germany

Abstract: Isomerization of the cyclic nitrone **N** into the corresponding lactam **L** and acetonitrile addition to nitrone **N** are enhanced by one-electron-oxidation of the substrate
and then proceed via radical cation chain reactions.

The sterically highly encumbered nitrone 5,7-di-tert-butyl-3,3-dimethyl-3H-indole-1-oxide
(**N**)[2] undergoes a facile isomerization at room temperature into its isomeric and ther-
modynamically more stable lactam **L** by interaction with

(a)   0.05 molar equivalents of tris(4-bromophenyl)aminylium hexachloroantimonate
      in methylene chloride solution,

(b)   photoexcited 9,10-dicyanoanthracene (**DCA**) in acetonitrile, preferably in the
      presence of biphenyl (**BP**).

In case (a), a trivial acid mediated isomerization [1] is ruled out from the fact that the
isomerization **A → L** ceases upon interruption of the cation salt addition.

In case (b) the isomerization **A → L** is paralleled by the addition of solvent acetonitrile
(analogous to addition of other nitriles via their CN triple bonds [1]) with formation of
an adduct **A**, presumably via its precursor **A'**. The relative importance of both pathways
is dependent on the concentrations of both **N** and **BP**. With constant $[DCA]$ = 3 x 10^{-4} M
and 400-440 nm irradiation, quantum yields of starting material decomposition (ϕ_N) and
product formation (ϕ_L, ϕ_A) run as follows: At $[BP]$ = 0.10 M and $[N]$ varying from

H. Fischer, H. Heimgartner (Eds.)
Organic Free Radicals
© Springer-Verlag Berlin Heidelberg 1988

0.003 M to 0.20 M ϕ_N runs from 0.5 to 3.3, ϕ_L from 0.04 to 1.7, and ϕ_A from 0.5 to 1.4. At $[N]$ = 0.03 M and $[BP]$ varying from zero to 2.50 M (solubility limit), ϕ_A increases from 0.6 to 5.9. Quantum yields larger than unity are an indication of chain reactions. Such processes, which either form two molecules of L out of two molecules of N or one molecule of A out of one molecule of N per cycle are envisaged as follows:

Start: $N + Ar_3N^{\ddot{+}} SbCl_6^- \longrightarrow N^{\ddot{+}} SbCl_6^- + Ar_3N$ (Ar = 4-Br-C_6H_4)

or: $DCA^* + BP \longrightarrow DCA^{\ddot{-}} + BP^{\ddot{+}}$

$BP^{\ddot{+}} + N \longrightarrow BP + N^{\ddot{+}}$

or: $DCA^* + N \longrightarrow DCA^{\ddot{-}} + N^{\ddot{+}}$

Chain process 1 (2 N → 2 L):

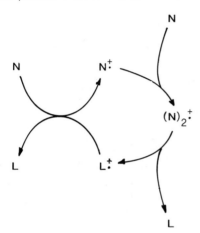

Chain process 2 (N → A):

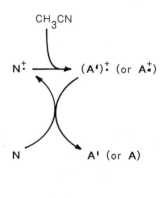

Formation of A' from N and acetonitrile formally constitutes a 1,3-dipolar cycloaddition. Since the formation of A is very slow at room temperature, the fast appearance of A and the high quantum yields of its formation upon DCA-sensitization would require that the addition of acetonitrile to N in this case is electron transfer mediated as depicted in chain process 2.

Participation of a dimer radical cation $(N)_2^{\ddot{+}}$ in process 1 is made likely on the basis of double labelling experiments: the mass spectrometrically determined isotopic composition (^{15}N, ^{18}O) of product L formed from a mixture of ^{18}O-N and ^{15}N-N is that of a statistical label distribution and clearly deviates from that of L isolated after combining two separate runs of isomerization of ^{15}N-N and ^{18}O-N.

References:

1 Taken in part from: Kretz H (1987) PhD thesis Univ. Duisburg.
2 Döpp D (1976) Chem. Ber. 109: 3849

Tetramethyleneethanes: Non-Kekulé
Ground State Triplets at Odds with Theory

Paul Dowd

Department of Chemistry, University of Pittsburgh
Pittsburgh, PA 15260 USA

The non-Kekulé molecules trimethylenemethane **1** and tetramethyleneethane **2** have degenerate non-bonding molecular orbitals. Both molecules are expected to be

1   **2**

ground state triplets in accord with Hund's rule. The situation with respect to trimethylenemethane is settled; theory and experiment are in good agreement, it is a ground state triplet.[1]

Tetramethyleneethane is another matter. Recent theoretical analysis[2] indicates that tetramethyleneethane is, like cyclobutadiene, a disjoint π system. Disjoint molecules are those in which a pair of degenerate non-bonding orbitals can be found which do not overlap spatially. It has been suggested[2] that electron correlation in **2** is satisfied by this criterion, and that the ground state of **2** need not be a triplet. Tetramethyleneethane **2** has been predicted on these grounds to be a ground state singlet.[2]

Recent research is aimed at a resolution of the conflict between theory and experiment using tetramethyleneethanes of various geometric configurations. All the hydrocarbon tetramethyleneethanes **2, 3, 4,** and **5** are *ground state triplets*, contrary to current theoretical expectation.[3,4] By contrast, the dimethylenefuran **6** is a ground state singlet. The latter observation has been taken as support for the theoretical position.[5]

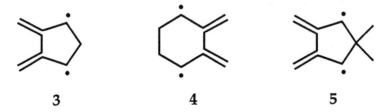

3   **4**   **5**

H. Fischer, H. Heimgartner (Eds.)
Organic Free Radicals
© Springer-Verlag Berlin Heidelberg 1988

6 a   **6 b**

An understanding of the difference in behavior between the singlet **6** and the triplets **2, 3, 4**, and **5** demands recognition that in the hydrocarbon tetramethyleneethanes there exist no opportunity for covalent bonding. This is the essence of the non-Kekulé designation. Molecules such as **6** and cyclobutadiene are not non-Kekulé molecules and adopt the singlet state configuration where interaction between the erstwhile radical centers leads to covalent bond formation.

Acknowledgement This research was generously supported by the National Science Foundation. It was a great pleasure to collaborate on this research with Yi Hyon Paik and Wonghil Chang.

References

(1) Baseman, R. J.; Pratt, D. W.; Chow, M.; Dowd, P. *J. Am. Chem. Soc.* **1976**, *98*, 5726; Platz, M. S.; McBride, J. M.; Little, R. D.; Harrison, J. J.; Shaw, A.; Potter, S. E.; Berson, J. A. *J. Am. Chem. Soc.* **1976**, *98*, 5725.

(2) Borden, W. T.; Davidson, E. R. *J. Am. Chem. Soc.* **1977**, *99*, 4587; Borden, W. T. in *Diradicals*; Borden, W. T., Ed.; Wiley: New York, **1982**, pp 1-72.

(3) Dowd, P. *J. Am. Chem. Soc.* **1970**, *92*, Dowd, P.; Chang, W.; Paik, Y. H. *J. Am. Chem. Soc.* **1986**, *108*, 7416; Dowd, P.; Chang, W.; Paik, Y. H. *J. Am. Chem. Soc.* **1987**, *109*, 5284.

(4) Roth, W. R.; Kowalczik, U.; Maier, G.; Reisenauer, H. P.; Sustmann, R.; Müller, W. *Angew. Chem. Int. Ed. Engl.* **1987**, *26*, 1285; Roth, W. R.; Erker, G. *Angew. Chem. Int. Ed. Engl.* **1973**, *12*, 503.

(5) Stone, K. J.; Greenberg, M. M.; Goodman, J. L.; Peters, K. S.; Berson, J. A. *J. Am. Chem. Soc.* **1986**, *108*, 8088; Zilm, K. W.; Merrill, R. A.; Greenberg, M. M.; Berson, J. A. *J. Am. Chem. Soc.* **1987**, *109*, 1567.

HYDROPHOBIC RADICAL-IONS

I. Drăguţan,[a] A. Caragheorgheopol,[b] H. Căldăraru[b] and R.J. Mehlhorn[c]

[a]Center of Organic Chemistry, R-78100 Bucharest, Romania
[b]Center of Physical Chemistry, Bucharest, Romania
[c]LBL, University of California, Berkeley, USA

Abstract: The synthesis of several cationic stable free radicals
with different positive charge characteristics, obtained by the
reaction of spin labelled primary amines with trisubstituted pyrylium perchlorates, is presented.

Recently we reported (1) the synthesis of nitroxides belonging to a
new class of cationic stable free radicals ($\underline{3}$) arising from the reaction of a spin-labelled primary amine (e.g. $\underline{1}$) and a 2,4,6-trisub-
stituted pyrylium perchlorate $\underline{2}$ (2).

$$\underline{1} \qquad\qquad \underline{2} \qquad\qquad\qquad\qquad \underline{3}$$

Additional radicals in this group possessing different positive
charge characteristics have been prepared by the same general method
choosing the substituents in the pyrylium perchlorate $\underline{2}$ so as to en-
sure increasing charge delocalization over one or several aromatic
rings (see Table). Such structures are expected to be suitable for
studies of membrane bioenergetics focusing on measurements of electrical potentials; these require ionic probes having significant rates
of membrane permeation which accumulate inside the cell in response
to electrical potentials, e.g. light-dependent potentials across pho-
toactive membranes.

Spin label $\underline{3b}$, prepared from 2,6-dimethyl-3,5-nonamethylenepyrylium
perchlorate (3), was synthesized with the consideration that the
hydrophobic nine-carbon methylene chain attached to the ring with delocalized positive charge should facilitate binding of the ion to
membranes and, possibly, impair facile ion-pair formation and reduce
the likelihood of diffusion of this cation across membranes in con-
junction with anions.

H. Fischer, H. Heimgartner (Eds.)
Organic Free Radicals
© Springer-Verlag Berlin Heidelberg 1988

Table 1 - Labelled pyridinium perchlorates $\underline{3}$

Compd.[a]	R_1	R_2	R_3	Reaction conditions			Yield[b] %	M.p. °C	Color
				ml EtOH g of $\underline{2}$	Time min.	Temp. °C			
$\underline{3a}$	Me	Me	Me	14	15	25	54	219	Yellow
$\underline{3b}$	Me	$(CH_2)_9$	Me	60	90	78	49	123	Pink-orange
$\underline{3c}$	Me	Ph	Me	80	30	78	57	242	Yellow
$\underline{3d}$	Me	⟨⟩–⟨⟩	Me	100	45	78	61	247	Light-pink
$\underline{3e}$	Ph	Ph	Ph	30	over-night	25	65	143	Orange

[a]Correct elemental analyses were obtained for all new compounds;
[b]Crude yields; [c]Ref. (1).

Preliminary resuslts obtained in studies (4) with new radical $\underline{3c}$ in envelope vesicle membranes of Halobacterium halobium indicate that it exhibits a higher permeability than the less delocalized $\underline{3a}$ or the quaternary ammonium nitroxides (5) derived from 2,2,6,6-tetramethyl-4-aminopiperidine-N-oxyl with point positive charges.

Acknowledgement: The authors are grateful to Prof. A.T. Balaban for the generous gift of samples of pyrylium perchlorates $\underline{2}$ (Ref. 2, 3).

1. I. Drăguţan, A.T. Balaban (1982) Can. J. Chem. 60, 1512.
2. a. A.T. Balaban, W. Schroth, G.W. Fisher (1969) Adv. Heterocycl. Chem. 10, 241; b. A.T. Balaban, A. Dinculescu, G.N. Dorofeenko, G.W. Fischer, A.V. Koblik, V.V. Mezheritskii, W. Schroth (1982) Adv. Heterocycl. Chem. Suppl. 2, Academic Press, New York.
3. A.T. Balaban (1968) Tetrahedron Lett. 4643.
4. R.J. Mehlhorn, L. Packer, R. Macey, A.T. Balaban, I. Drăguţan (1986) Meth. Enzymol. 127, 738.
5. A.T. Quintanilha, R.J. Mehlhorn (1978) FEBS Lett. 91, 104; R.J. Mehlhorn, L. Packer (1979) Meth. Enzymol. 56, 515.

SIMPLE SYNTHESIS OF NITROXYL AMINES

I. Drăguţan[x], V. Drăguţan[x] and R.J. Mehlhorn[xx]

[x]Center of Organic Chemistry, R-78100 Bucharest, Romania
[xx]LBL, University of California, Berkeley, CA 94720, USA

Abstract: A one-step procedure of general use for the synthesis of spin labelled amines in high yield is described.

Spin labelled amines find extensive applications as versatile label-ling reagents (1), for studies of spin-metal interactions (2) or for measurement of pH gradients across biological membranes (3). The availability of different labelled amines provides the opportunity for the most appropriate selection of pK_a, reactivity and distance between the amino and paramagnetic groups for specific biological studies. To achieve optimization of such studies considerable ef-fort has been devoted to the synthesis of nitroxyl amines (4).

Hideg et al.(5) contributed to the series of previously known amino-methyl nitroxides 2 (6) and 3 (7) by obtaining the allylic amine 1; the latter was prepared from the mesylate 4 ($X = OSO_2CH_3$) by a reac-

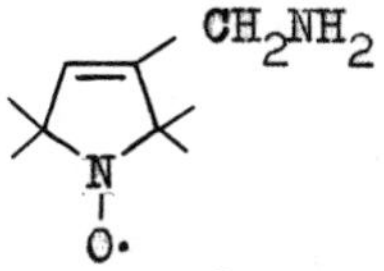

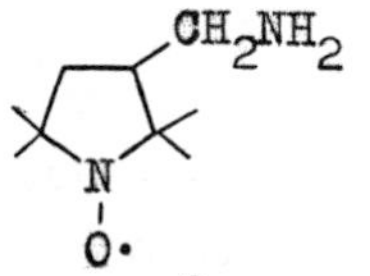

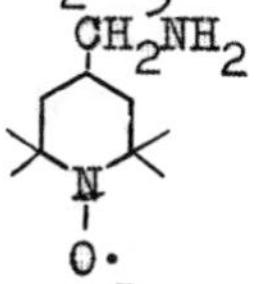

tion sequence involving the corresponding azide and triphenylphos-phinimine and was isolated as its tosylate in 64% yield.

We now report a shorter, one-step procedure starting from 4 ($X = Br$ or $OSO_2C_6H_4CH_3$) (8) which on treatment with NH_3–MeOH directly gave pure amine 1 in 75-80% yield. The method has been successfully used for preparing new parent secondary amines 5 by reacting 4 with the appropriate primary amine as illustrated by the synthesis of 3-(N-t-butylaminomethyl)-2,2,5,5-tetramethyl-2,5-dihydropyrrole-1-oxyl.

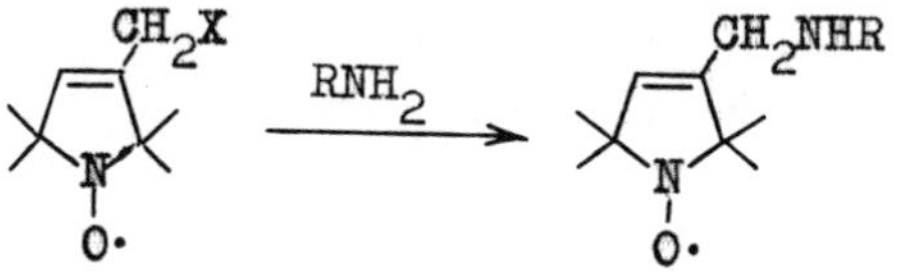

H. Fischer, H. Heimgartner (Eds.)
Organic Free Radicals
© Springer-Verlag Berlin Heidelberg 1988

EXPERIMENTAL

3-Aminomethyl-2,2,5,5-tetramethyl-2,5-dihydropyrrole-1-oxyl (1)

Crude $\underline{4}$, X = $OSO_2C_6H_4CH_3$ (1 g; 3.08 mmol) in 5 ml anhydrous methanol was added dropwise over 15 min. into 50 ml of methanol saturated with dry gaseous ammonia. Stirring was continued for 2 hr at room temperature, then the solution was left to stand overnight. Evaporation to dryness yielded a residue which was treated with 40 ml of buffer solution pH 5 (citric acid - Na_2HPO_4), repeatedly extracted with ether to remove impurities in $\underline{4}$, then the solution was saturated with sodium hydroxide (pellets) and extracted four times with ethyl ether; the ethereal extracts were dried over potassium carbonate and concentrated to give 0.42 g of a red oil. IR (CH_2Cl_2): ν_{NH} 3330 and 3430 cm^{-1}; the ESR spectrum (in benzene) yields a triplet with a_N = 14.4 G and g = 2.00568. ^{1}H-NMR ($CDCl_3$) after phenylhydrazine reduction gave: 1.28 (s, 6H) and 1.4 (s, 6H) for the four ring methyls, 1.17 (s,2H)(NH_2, drifting with dilution), 3.2 (d, 2H, $C\underline{H}_2NH_2$; J=1Hz) and 5.35 ppm (t, 1H, J = 1 Hz, olefinic proton). Anal. Calcd. for $C_9H_{17}N_2O$: C, 64.03; H, 10.08; N, 16.55. Found: C, 63.92; H, 10.23; N, 16.65%.

3-(N-t-Butylaminomethyl)-2,2,5,5-tetramethyl-2,5-dihydropyrrole-1-oxyl (5, R = t-butyl)

To a solution of 0.6 g (1.85 mmol) tosylate $\underline{4}$ in 4 ml of methanol 1 ml (9.5 mmol) $\underline{t}$-butylamine was added and the reaction mixture was left to stand for two days at room temperature. The same work-up as above yielded 70% $\underline{5}$, R = $\underline{t}$-Bu, yellow crystals m.p. 57-59°C(purified by sublimation on a steam bath, in vacuo); ESR (benzene): a_N =14.51G, g = 2.0058.

1. a. B.J. Gaffney (1976) In: L.J. Berliner (ed) Spin Labeling. Theory and Applications. Academic Press, New York; b.B.K. Sinha, M.G. Cox, C.F. Chignell, R.L. Cysyk (1979) J. Med. Chem. 22, 1051; c. R.I. Zhdanov, N.M. Mirsalikhova, E.G. Rozantsev, B.V. Rozynov, O.S. Reshetova (1979) Bioorg. Khim. 5, 1385.
2. J.L. Dreyer, H. Beinert, J.F.W. Keana, O. Hankovszky, K. Hideg, S.S. Eaton, G.R. Eaton (1983) Biochem. Biophys. Acta 745, 229.
3. R.J. Mehlhorn, L. Packer (1983) Ann. New York Acad. Sci. 180.
4. a. E.G. Rozantsev (1970) Free Nitroxyl Radicals, Plenum Press,N.Y. b. G.M. Rosen (1974) J. Med. Chem. 17, 358; c. J. Pirrwitz, W.Damerau (1976) Z. Chem. 16, 401; d. R.I. Zhdanov, Z.L. Gordon, E.G. Rozantsev (1975) Dokl. Akad. Nauk 224, 593; e. R.I.Zhdanov, E.G. Rozantsev (1980) Izv. Akad. Nauk SSSR Ser. Khim. 364.
5. H.O. Hankovszky, K. Hideg, L. Lex (1981) Synthesis 147.
6. J.C. Hsia, L.H. Piette (1969) Arch. Biochem. Biophys. 129, 296.
7. A.B. Shapiro, L.S. Bogach, V.M. Chumakov, A.A. Kropacheva, V.I. Suskina, E.G. Rozantsev (1975) Izv. Akad. Nauk SSSR 2077.
8. H.O. Hankovszky, K. Hideg, L. Lex (1980) Synthesis 914.

MICELLAR AGGREGATION AND THE DECAY
OF TRIPLET RADICAL PAIRS.

C. Evans and J.C. Scaiano.

Division of Chemistry,
National Research Council of Canada,
Ottawa, Canada, K1A-0R6.

Abstract: Magnetic field effects (MFE) on micellized triplet radical or radical ion pairs (RP) involving Vitamin E and the botanical phototoxin ,α-terthienyl, have been observed. Micellar dimensions are found to influence the decay behaviour of both systems. Heavy atom substitution is found to reduce the MFE in the terthienyl case.

EFFECTS OF MICELLAR AGGREGATION ON A NEUTRAL RP:

We report here effects of changing the aggregation number of sodium-n-alkyl sulfate surfactants on the decay behaviour of the micellized Vitamin E phenoxyl-butyrophenone ketyl triplet RP. MFE on this RP, which is formed via H abstraction from the Vitamin E hydroxy group by the triplet ketone, have recently been reported (1). The kinetic parameters of concern are the geminate recombination (k_{gem}) and the escape (k-) rate constants which have been calculated as in reference 1.

Changes in salt concentration are known to influence micellar aggregation. At room temperature a plot of $1/k_{gem}$ vs. [NaCl] followed the same trend as a plot of SDS aggregation number vs. [NaCl] - showing upward curvature above 0.4M NaCl. i.e. k_{gem} becomes smaller as the micelle becomes larger. The variation of 1/k- with NaCl was in the same direction but was linear.

Temperature changes may also alter micellar dimensions. Arrhenius plots (temperature range 5 to 85C) based on k- in SDS and SDecylS were linear and gave the following activation parameters which were independent of chloride up to the highest concentration tested (0.5M): Ea=7.3±1.0 kcal/mol, logA=11.1±0.5 for SDS; Ea=5.8±0.3 kcal/mol, logA=10.3±0.2 for SDecylS.

Arrhenius plots (temperature range 5 to 85C) based on k_{gem} in SDS and SDecylS were found to be non-linear (Figure 1). This figure shows data for SDS only. Similar results were obtained with SDecylS.The inflection points did not correlate with the critical micelle temperature of either surfactant. This behavior can be explained qualitatively by reference to the variation in micellar radius with temperature reported in reference 2. Our results suggest that k_{gem}, which we belive is determined by intersystem crossing in the triplet RP, is a fairly sensitive probe of the dimensions of the micellar environment.

H. Fischer, H. Heimgartner (Eds.)
Organic Free Radicals
© Springer-Verlag Berlin Heidelberg 1988

MFE ON A RADICAL CATION PAIR:

The T_1 state of α-terthienyl (αT) transfers an electron to methyl viologen (MV) with $k=(6.8\pm0.3)\times10^9 M^{-1}s^{-1}$ in methanol leading to $\alpha T^{+\bullet}$ ($\lambda max=530nm$, $\varepsilon=29000M^{-1}cm^{-1}$

). The same reaction occur in SDS and we have observed a pronounced MFE on the triplet $\alpha T^{+\bullet}/MV^{+\bullet}$ pair. Typical results for αT in SDS are: $k_{gem}=4.7\times10^6 s^{-1}$: $k_-=0.31\times10^6 s^{-1}$; %escape at zero field=4.3%; %escape at 5.2kG=71%. Similar results were obtained for 2-cyano-αT in SDS. For the 2-bromo derivative $k_{gem}(=9.1\times10^6 s^{-1})$ is nearly twice as large as that for αT and the %escape is reduced to 22% at 5.2kG. Finally the ratio of surface areas of SDS/SOctylS is 2.6 (3). The observed ratio of k_{gem} in SOctylS to that in SDS is 2.5 and 2.4 for αT and CN-αT, respectively. This is consistent with our hypothesis that k_{gem} is a probe of micellar dimensions.

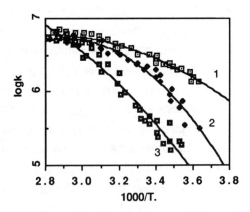

Figure 1: Arrhenius plots based on k_{gem} for the Vitamin E phenoxyl butyrophenone ketyl triplet radical pair in 0.15M SDS. [NaCl]: 0.0 (1); 0.25 (2); and 0.5M (3).

1 Evans C, Ingold KU, Scaiano JC (1988) J Phys Chem 92:1257
2 Mazer NA, Benedeck GB, Carey MC (1976) J Phys Chem 80:1075.
3 Missel PJ, Mazer NA, Benedeck GB, Carey MC (1983) J Phys Chem 87:1264.

SPIN TRAPPING AND MATRIX ISOLATION E.S.R. STUDIES OF PHOTO-OXIDATION
OF ALCOHOLS BY HEXACHLOROMETALLATE (IV) IONS (M = Pt, Pd, Ir)

Anand G Fadnis and Terence J Kemp

Department of Chemistry, Holkar Science College, Indore-452001 INDIA and

Department of Chemistry, University of Warwick, Coventry, CV4 7AL, UK

The reduction potentials of $[MCl_6]^{2-}$ ions (M = Pt, Pd, Ir) are enhanced on oxidation into the visible bands, and new photo-redox processes can be realised, e.g. oxidation of alcohols by $([PtCl_6]^{2-})*$ to yield, successively, Pt(II) and Pt metal. [Inorg.Chem.25, 2910, (1986)] We report the site of oxidation of alcohols (RCH_2OH, R_2CHOH, R'R"CHOH, $CH_2OH(CHOH)_n CH_2OH$) by $([MCl_6]^{2-})*$ using E.S.R. Technique of spin trapping and matrix isolation.

C-centred free radical intermediates formed during the mild photolysis (λ = 380nm) of hexachlorometallate(iv) ions in aqueous alcohol solutions were detected. Whereas in pure alcoholic media many of the oxidations lead to O-centred radicals. Sodium-2-sulphonato phenyl - t -butyl nitrone(SPBN) and α -phenyl -N-t-butyl nitrone(PBN) were used as spin traps.(fig.1)

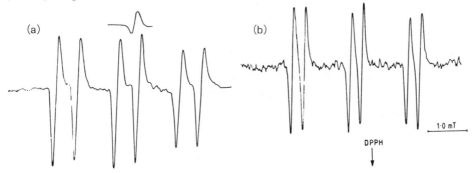

Fig.1 (a) E.s.r. spectrum of spin adduct of CH_2OH and SPBN obtained on photolysis of $[PtCl_6]^{2-}$ in MeOH - Water (a_N = 1.53mT, a_H = 0.51mT).

(b) E.s.r. specturm of spin adduct of CH_2O and PBN obtained on photolysis of $[PdCl_6]^{2-}$ in MeOH (a_N = 1.45mT, a_H = 0.29mT)

Prolonged, wide band photolysis (λ = 300nm) of $[PtCl_6]^{2-}$ solutions in aqueous MeOH and EtOH using SPBN as spin trap gave rise to a radical mixture consisting of the spin adducts of C-centred radical from alcohol and hydrogen atom possibly produced due to the appearance of colloidal Pt in the later stage of photolysis(fig.2).

H. Fischer, H. Heimgartner (Eds.)
Organic Free Radicals
© Springer-Verlag Berlin Heidelberg 1988

(a)

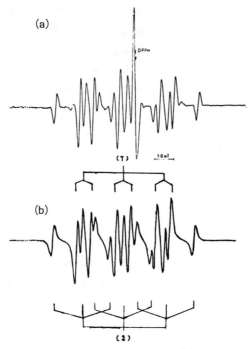

Fig.2 (a) E.s.r. spectra of the spin adducts of (1) CH$_2$OH, (2) H and SPBN obtained on photolysis of [PtCl$_6$]$^{2-}$ in MeOH water mixture. (b) Computer simulated spectra.

At 77K, wide band photolysis (λ = 300nm) of [MCl$_6$]$^{2-}$ (M = Pt, Pd) ions in alcohols and MeCN gave rise to Pt(III) and Pd(III) centres, identified by their g$\perp$ features in addition to the expected C-centred organic radical(fig.3).

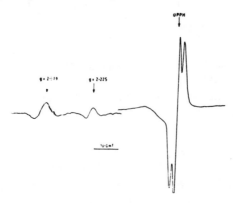

Fig.3 Second derivative e.s.r. spectrum obtained on photolysis of [PtCl$_6$]$^{2-}$ in CD$_3$OD at 77K

On the basis of these results both C-H and O-H abstraction processes were included in the proposed reaction mechanism of the reactions under study.

ESR OF FREE RADICALS IN GAMMA IRRADIATED DNA-HISTONE COMPLEX

A.Faucitano,A.Mele,A.Buttafava,F.Martinotti
Dip. di Chimica Generale V.le Taramelli 12 PAVIA (Italy)

Alexander and coll. have shown for the first time that the ESR spectrum of the nucleo proteins is essentially that of DNA and have interpreted this result in terms of energy transfer phenomena from the proteins to DNA (1).To similar conclusions arrived also S.C.Lillicrap , E.N.Felder (2) and M.Kuwabara et coll.(3) who also have extended the investigation to DNA-polylysine and DNA-polyalanine complexes.More recently M.C.R Symons and P.M.Cullis (4) have obtained esr evidence that the proteins act as radiosensitizers by increasing the yield of electron capture products,mainly the thymine radical anion.In this note we refer on the results of a quantitative esr study of the radical products of the gamma radiolysis of DNA-Histone complexes and DNA-Histone random molecular solid solutions.The comparison with the random solid solution is aimed to get informations on the role played by the structure of the nucleoprotein in the mechanism of radiosensitization.

Pure DNA from calf thymus was purified by treatement with the proteolytic enzyme pronasys ;a protein residue of of less than 0.4 % was determined by fluorimetry after the treatement .The complex DNA-Histone (type VI S Sigma) was prepared by the method of the dialysis gradient. The random solid solutions were prepared by evaporation of 1.8 M solutions of $HN(Et)_3HCO_3$ at pH 7.84 containing 60% Histone type and 40 % DNA.The ionic strength of such solutions was high enough to prevent the association DNA-Histone and the self aggregation of the proteins.The formation of the DNA-Histone complex was checked by electron microscopy.The irradiations were performed at R.T.under vacuum in a ^{60}Co gamma source with total doses of about 5 Mrad.The ESR spectra were recorded on a Varian E 109 spectrometer.

RESULTS

In fig 1 a,b the esr spectra of irradiated pure DNA and Histone are reported.In addition to the well known octet component due to the thymil radical(5) ,the spectrum of irradiated DNA is interpreted in term of the superimposition of a doublet of 23 G. and an unresolved central peack ,for which only tentative assignements can be made (6).The doublet of the Histone spectrum is attributed to the amido radical -NH-C(H)-C(O)-.The ESR spectrum of the mechanical mixture of DNA and Histone 1:1 by weight is simulated by combination of 78.5 % and 21.5 % of the DNA and Histone patterns respectively.This implies that the radiolytic radical yield of the protein is greater by a factor of 3.6.As compared to the mechanical mixture,the ESR spectrum of the molecular complex shows an increase of the DNA component to 48 % of the total area which is diagnostic of a transfer of about 30 % of the radiation damage from the protein to DNA(fig 1c,d).In the spectrum of the irradiated random solid solution the contribute of the DNA pattern shows a further increases to 80 % which can be reckoned to a transfer of about 60 % of the radiation damage(fig 1 e,f).According to the electron microscopy,the structure of the nucleohistone can be described as a sequence of agglomerates of proteins about 100 A x 25 A, surrounded by DNA super helics.The agglomerates are connected by segments of uncomplexed DNA.This configuration produces a relatively poor molecular contact between DNA and the protein and it must involve extensive migration of the active species through the protein moiety for the

H. Fischer, H. Heimgartner (Eds.)
Organic Free Radicals
© Springer-Verlag Berlin Heidelberg 1988

transfer to occur.A different situation applies in the case of the random molecular mixture:here the higher efficiency of transfer for the radiation damage can be reckoned with the more intimate contact between DNA and the protein which is made possible by the inhibition of protein self aggregation by the solution high ionic strength.In this latter system presumably a greater frequency of hydrogen bonds and overlaps is available which may act as bridges for the transfer of energy, electrons and free valencies.An important conclusion stemming from the computer simulations is that the enhancement of the DNA pattern is essentially homogeneous involving all the h.f. conponents.This implies that not only the radiolytic yield of the thymil radicals and their precursors radicals anions is enhanced but that also the species contributing to the singlet and the doublet take part in the energy transfer process.Such h.f.components have not yet conclusively identified but I.N.D.O. M.O calculations suggest that the doublet may arise from the H atom addition at N_7 of G and A,whilst the singlet is compatible with the H addition at O_4 of T,O_6 of G and O_2 of C(6).As in the case of Thymil radicals,the additions of H atoms may occurr either directly or by electron transfers followed by protonation of the anion radicals.

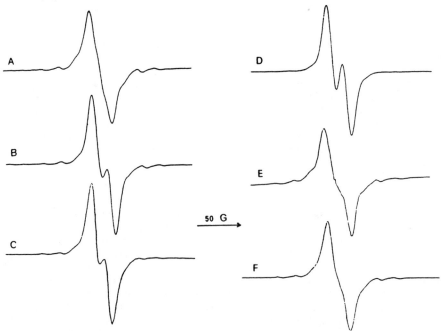

Fig. 1 ESR spectra obtained by gamma irradiation of DNA-Histone samples:
          A,pure dry DNA; B,pure dry hystone; C, DNA-Histone complex ;
D,computer simulation of C; E, DNA-Histone random solid solution;F,
computer simulation of E

References
1 P.Alexander,J.T.Lett,M.G.Ormerod Biochim.Biophys Acta 51 ,207,(1961)
2 S.C.Lillicrap,E.M.Fielden Int.J.Rad.Biol. 21,137,(1972)
3 M.Kuwabara,M.Hayashi,G.Yoshii J.Rad.Res. 14,198,(1973)
4 P.M.Cullis,M.C.R.Symons,M.C.Sweeney,G.D.D.Jones,J.D.McClymont
  Proceedings Intern. Congress,of Radiation Research EDIMBURG 1987 Pg 12
5 S.Gregoli,A.Bertinchamps Int.J.Radiat.Biol. 21,65,(1972)
6 A.Faucitano,A.Mele,A.Buttafava,F.Martinotti J.Chem.Soc. P II 329,(1985)

AN ESR INVESTIGATION OF THE
PHOTOLYSIS OF BENZENE
POLYCARBOXYLATE ANIONS IN AQUEOUS SOLUTION

R. W. Fessenden and A. S. Jeevarajan

Radiation Laboratory and Department of Chemistry,
University of Notre Dame, Notre Dame, IN 46556, U. S. A.

The radicals produced by continuous uv photolysis of aqueous solutions of the anions of benzene polycarboxylic acids have been
detected and identified by means of their ESR spectra. The behavior
of terephthalate (1,4-benzenedicarboxylate) under basic conditions
(pH 13) is typical. With terephthalate alone, only the lines of the
trianion radical are seen. In the presence of 1.3 M 2-propanol, the
intensity of these lines is greatly enhanced and joined by those of a
new radical, which can be identified from its spectrum to be the
cyclohexadienyl radical formed by H addition at a ring carbon atom
bearing a carboxyl group. The parameters of this radical (in gauss)
are $a(CHCO_2^-)=42.93$, $a(ortho)=8.15$, $a(meta)=2.34$ G and g factor
2.00305. Addition of acetone further enhances the intensities of the
lines of these radicals. The chemical mechanism proposed to account
for the radical formation is quenching of the triplet state of terephthalate by the alcohol with competitive H transfers to a carboxyl-
ate oxygen and to the substituted ring position. Optical laser photolysis studies support this mechanism in that triplet state quenching by the alcohol is observed along with corresponding formation of
the trianion radical [1]. Analogous ESR results were found with many
of the other benzene carboxylates up to the hexacarboxylate. (With
phthalate only the anion is seen.) In the two cases where the carboxyl groups are not opposite each other (1,3-benzenedicarboxylate
and 1,3,5-benzenetricarboxylate) H addition occurs opposite one of
the carboxyl groups. In 1,2,4-benzenetricarboxylate two isomers are
seen with addition at either 1 or 4 positions.

The spectra of the cyclohexadienyl radical from terephthalate is
notable in that the low-field lines appear in emission and the
enhancement factor can be > 10 depending on the rate of production.
Emission is seen even for the corresponding radicals from the more

H. Fischer, H. Heimgartner (Eds.)
Organic Free Radicals
© Springer-Verlag Berlin Heidelberg 1988

highly substituted (and charged) examples which clearly cannot be reacting by bimolecular reaction at a rate fast enough to produce such a large polarization. Time resolution provided by chopping the light with a sector shows that the majority of the polarization originates in the radical formation step. We suggest that the polarization is produced by the radical pair mechanism at the separation of the cyclohexadienyl and 2-propyl-2-ol radicals rather than during radical disappearance reactions. The long relaxation time of 8 μs measured for the radical from terephthalate enhances the observed polarization.

1 Weir D   To be published

THE EFFECT OF TEMPERATURE IN THE ADDITION
OF TOSYL RADICALS TO ARYLACETYLENES

Carlos M. M. da Silva Corrêa and Maria Daniela C. M. Fleming

Centro de Investigação em Química (INIC), Faculdade de Ciências
4000 PORTO – Portugal

INTRODUCTION:

The effect of substituents and temperature on the reactivity of arylacetylenes toward tosyl radicals generated by photolysis of tosyl iodide was studied. Relative reactivities yield good Hammett-Brown correlations. Relative Arrhenius parameters were determined and its effects on the reactivity analysed.

SELECTIVITIES:

The relative reactivities of $X-C_6H_4C{\equiv}CH$ (X = p-MeO, p-Me, H, p-Cl, m-NO$_2$) towards p-Me-$C_6H_4SO_2^{\bullet}$ were measured by competition experiments, at several temperatures (T/ °C = -25, 0, 25, 50, 70), by using equation (3):

$$p\text{-Me-}C_6H_4SO_2^{\bullet} + X-C_6H_4C{\equiv}CH \longrightarrow \text{Radical adduct } \{k_X\} \quad (1)$$

$$p\text{-Me-}C_6H_4SO_2^{\bullet} + C_6H_5C{\equiv}CH \longrightarrow \text{Radical adduct } \{k_H\} \quad (2)$$

$$k_X/k_H = (\log[XA] - \log[XA]_o)/(\log[A] - \log[A]_o) \quad (3)$$

where [A] and [XA] are the concentrations of both acetylenes in competition.[1] Plots of $\log(k_X/k_H)$ *versus* σ^+ gave good correlations showing the importance of polar effects. At all the temperatures examined, phenylacetylenes with electron-donating substituents react faster. Results of competition experiments at several temperatures are summarized in **Table 1**. A plot of the ρ^+ values against $1/T$ gave a concave upward curve with the minimum near 25 °C.

ACTIVATION PARAMETERS:

Relative Arrhenius parameters have been determined from plots of $\log(k_X/k_H)$ *versus* $1/T$. Results are given in **Table 2** and show that all the substituents decrease the pre-exponential factor; the retardation ability, how far as $\Delta S^{\ddagger}$ is concerned, is: m-NO$_2$ > p-Cl > p-Me > p-MeO > H. In the respect of $\Delta H^{\ddagger}$, all the substituents increase the reactivity: p-MeO > m-NO$_2$ > p-Cl > p-Me > H.

H. Fischer, H. Heimgartner (Eds.)
Organic Free Radicals
© Springer-Verlag Berlin Heidelberg 1988

No correlation between $\log(A_X/A_H)$ and σ^+ or between ΔE_a and σ^+ could be obtained.

Table 1–Relative reactivities, ρ^+ values, and goodness of fit for the addition of $p\text{-Me-C}_6H_4SO_2^\bullet$ to $X\text{-C}_6H_4C\equiv CH$ in CCl_4, at several temperatures.

X	Temperature / ^{0}C				
	-25	0	25	50	70
p-MeO	3.15	3.00	2.83	2.46	2.10
p-Me	1.64	1.55	1.44	1.42	1.32
H	(1)	(1)	(1)	(1)	(1)
p-Cl	1.25	1.01	0.94	0.97	0.92
m-NO$_2$	0.67	0.55	0.48	0.48	0.49
ρ^+	-0.46	-0.51	-0.53	-0.48	-0.43
r	0.97	0.99	1.00	1.00	1.00
$\pm t \cdot s_\rho$(90%)	0.13	0.08	0.05	0.04	0.03
$\pm t \cdot s_\rho$(95%)	0.17	0.10	0.06	0.05	0.04

Table 2–Relative Arrhenius parameters and goodness of fit for the addition of $p\text{-Me-C}_6H_4SO_2^\bullet$ to $X\text{-C}_6H_4C\equiv CH$.

X:	p-MeO	p-Me	H	p-Cl	m-NO$_2$
$\ln(A_X/A_H)$:	-0.199	-0.268		-0.832	-1.630
$s_{x/y}$:	0.27	0.08		0.26	0.28
$\pm t \cdot s_{x/y}$(90%):	0.54	0.15		0.52	0.57
$\pm t \cdot s_{x/y}$(95%):	0.69	0.20		0.66	0.73
A_X/A_H:	**0.82**	**0.77**	**1**	**0.44**	**0.20**
k_H/k_X (if $\Delta E_a = 0$):	1.22	1.30		2.27	5.00
$E_{aH}-E_{aX}$ / kcal:	**0.69**	**0.37**	**0**	**0.49**	**0.58**
$s_{x/y}$:	0.16	0.04		0.15	0.16
$\pm t \cdot s_{x/y}$(90%):	0.32	0.09		0.30	0.33
$\pm t \cdot s_{x/y}$(95%):	0.41	0.11		0.38	0.42
k_X/k_H (if $A_X/A_H = 1$):	3.21	1.87		2.29	2.66
r:	0.93	0.98		0.88	0.90
Overall effect :	**E$_a$**	**E$_a$**		**E$_a$, A**	**A**

[1] C. M. M. da Silva Corrêa and M. D. C. M. Fleming (1987) *J. Chem. Soc., Perkin Trans. 2*, 103.

CHEMICAL ROUTES TO ATOMIC NITROGEN

A.R. Forrester, H. Irikawa, M. Passway,
R. Ritchie and K. Tucker

Chemistry Department, University of Aberdeen,
Old Aberdeen AB9 2UE, Scotland

The intensive investigation of both monovalent (RN) and divalent (R_2N) nitrogen intermediates during the last 25 years has been the source of much new chemistry. By comparison atomic nitrogen has received little attention and we are aware of only a few studies [1] of its reactions in solution. In these it was generated mainly in the ground state (quartet) by passing molecular nitrogen through a microwave discharge tube. We know of no solution reaction which will yield atomic nitrogen in a more controlled and less energetic way (for $N{\equiv}N\ D^0_{298} \simeq 980\ kJ\ mol^{-1}$) suitable for use in chemical studies. In contrast chemical generation of trivalent carbon (RC carbynes) [2] and active carbon [3] have been reported.

What is required is a readily available, stable substrate which will yield this unique reactive intermediate on mild thermolysis or photolysis so that subsequent investigation of its chemical reactivity will not be hampered by the conditions used in its production. Formation of atomic nitrogen from a diamagnetic precursor requires cleavage of three bonds most simply accomplished in two steps, viz two bond cleavage followed by one bond cleavage (Route 1) or <u>vice versa</u> (Route 2). In both cases it is essential that either the

ROUTE 1

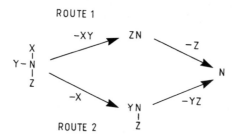

second step follows on rapidly from the first before the initially formed nitrene (ZN) or nitrogen radical (YZṄ) can react in some other way or even better that the two steps are concerted.

H. Fischer, H. Heimgartner (Eds.)
Organic Free Radicals
© Springer-Verlag Berlin Heidelberg 1988

With this aim a number of heterocycles (1-5) have been synthesised and suitably derivatised, X=Cl, Br, NO, CO_3But such that a driving force for three-bond cleavage at nitrogen is created. Using a combination of product determination, trapping and direct

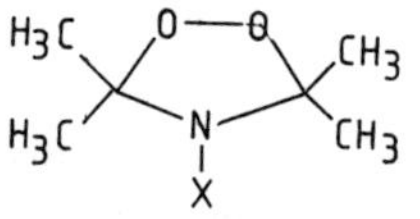

(1) 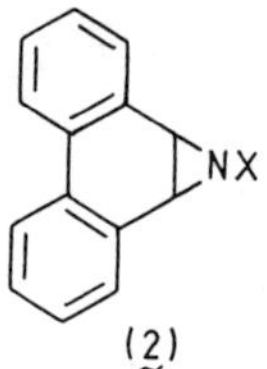   (2) 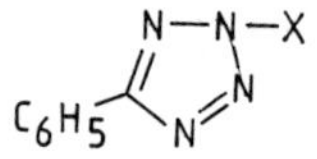   (3)

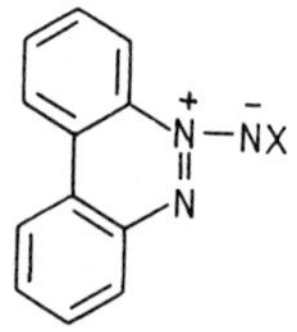

(4) 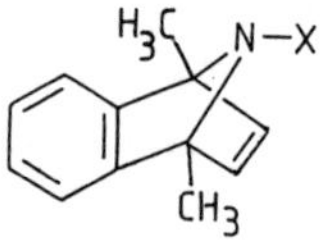   (5)

e.s.r. measurements evidence has been sought for the production of atomic nitrogen and its short-lived precursors from these substrates on mild thermolysis and/or photolysis.   A full description of this work will be presented.

1   Takahashi S, Miyazaki S (1977) Bull Chem Soc Japan 50: 1627; Froben FW (1978) Ber Bunsenges Phys Chem 82: 9; Havel JJ, Shell PS (1972) J Org Chem 37: 3744
2   Strausz OP, Kinnepohl GJA, Garneau FX, Dominih T, Kim B, Valenty S, Shell PS (1974) J Am Chem Soc 96: 5723
3   Shevlin PB, Kamnula S (1977) J Am Chem Soc 99: 2627

HYDROGEN ATOM ABSTRACTION FROM A MOLYBDENUM HYDRIDE: RADICAL CLOCK STUDIES[1]

J.A. Franz, J.C. Linehan, and M.S. Alnajjar

Pacific Northwest Laboratory
P. O. Box 999, Richland, Washington 99352 USA

Abstract: Rate expressions for the reaction of primary, secondary, and tertiary radical clocks with the molybdenum hydride $(CH_3)_5CpMo(CO)_3H$ are reported. Relative rates of group displacement reactions from primary, secondary and tertiary alkyl bromides, phenyalkyl selenides and sulfides by a molybdenum radical are presented.

INTRODUCTION

Accurate rate expressions for the reaction of alkyl radicals with metal hydrides provide important mechanistic tools for determining rates of rearrangement, electron transfer, and atom transfer rates involving metal systems.[2,3] Thus, we present results of a study of the reactions of radical clocks hex-5-enyl, hept-6-en-2-yl, and 2-methylhept-en-2-yl with $(CH_3)_5CpMo(CO)_3H$. The selectivity of the 17-electron molybdenum radical in reactions with primary, secondary and tertiary alkyl bromides, and phenylalkyl sulfides and selenides is presented.

EXPERIMENTAL

Organic perester or bromide precursors were used to generate alkyl clock radicals in dodecane. Photolysis and/or thermolysis of radical precursors in the presence of $(CH_3)_5CpMo(CO)_3H$ gave unrearranged, reduced hydrocarbon product (E) or rearranged, reduced hydrocarbon (D), whose product ratios gave relative rates of cyclization/abstraction (kre/kabs) of the unrearranged radical according to r = kre/kabs = [D][DH]$_{av}$/[E] for short extent of consumption of hydride [DH]$_{av}$ = average hydride concentration, or by computer solution of eq 1, for more extensive consumption of hydride, where B_0 is the initial concentration of hydride, f = 1 for alkyl halides and f = 2 for peresters, which consume two equivalents of hydride per mole of alkane formed:[4]

$$E + D = (1/f)(B_0 + r)(1 - e^{-fD/r}) \qquad (1)$$

Published rate expressions[5,6] of the three radical clock rearrangements were used to convert values of kre/kabs to kabs. Results are given in Table I. Relative rates of reaction with $(CH_3)_5CpMo(CO)_3H$ with a number of reagents were determined in dodecane. The results are presented in Table II.

H. Fischer, H. Heimgartner (Eds.)
Organic Free Radicals
© Springer-Verlag Berlin Heidelberg 1988

DISCUSSION

Results of Table I show a distinctive steric retardation of the tertiary radical attack at the crowded molybdenum not observed for <u>less</u> exothermic reactions of primary, secondary, and tertiary radicals with Bu_3SnH[6] and $PhSH$[7]. The overall rate of reaction of this hydride with halides is slower than Bu_3SnH because of a sluggish halogen atom abstraction step, but sufficiently fast and selective (compare Bu_3SnH selectivity[8]) for synthetically useful reductions of 1°, 2° and 3° bromides.

Table I. Arrhenius Expressions for the Reaction of Primary, Secondary and Tertiary Radical Clocks with $(CH_3)_5CpMo(CO)_3H$ in Dodecane.

Radical	$\log(A/M^{-1}s^{-1})$	$Ea/kcal/mol$	$k/M^{-1}s^{-1}$, 298 K
Hex-5-enyl	9.27 ± .04	1.35 ± .11	1.9×10^8
Hept-6-en-2-yl	8.66 ± .06	0.96 ± .43	9.1×10^7
2-Methyl-hept-6-en-2-yl	9.46 ± .04	3.41 ± .13	9.2×10^6

Table II. Relative Rates of Reaction of Sulfides, Selenides, and Bromides with $(CH_3)_5CpMo(CO)_3H$ in Dodecane at 100° C.

Radical Precursor	Relative Rate
$CH_2=CH(CH_2)_4SPh$	(1)
$CH_2=CH_2(CH_2)_3CH(SePh)CH_3$	15
$CH_3(CH_2)_7Br$	1000
$CH_2=CH(CH_2)_4Br$	1000
$c-C_6H_{11}Br$	1700
$CH_2=CH(CH_2)_3C(Br)(CH_3)_2$	35000

REFERENCES

1This work was supported by the Office of Basic Energy Sciences, U. S. Department of Energy, under Contract DE-AC06-76RLO 1830
2 Bullock RM, and Samsel EG (1987) J Am Chem Soc 109:6542-6544
3 Ash CE, Hurd PW, Darensbourg MY, Newcomb M., (1987) J Am Chem Soc, 109:3313-3317
4 Rüchardt C (1961) Chem Ber 94:2599
5 Chatgilialoglu C, Ingold KU and Scaiano JC (1981) J Am Chem Soc 103:7739-7742
6 Lusztyk J, Maillard B, Deycard S, Lindsay DA, and Ingold KU (1987) J Org Chem 52:3509-3514
7 Franz JA, Bushaw BA, and Alnajjar MS (1988) J Am Chem Soc, submitted
8 Beckwith ALJ, Pigou PE (1986) Aust J Chem, 39:72-87

RADICAL CYCLIZATION OF ALDEHYDES:
SYNTHETIC AND MECHANISTIC DEVELOPMENTS

B. Fraser-Reid

Paul M. Gross Chemical Laboratory, Department of Chemistry
Duke University, Durham, North Carolina, 27706 USA

Abstract: The material to be presented in this lecture grew out of the recent observation, summarized in Scheme 1, which shows that the highly-branched compound 1, on treatment with tri-n-butyltin hydride, gave a 4:1 mixture of the cyclohexanol 2 and the diquinane 3 [1]. Subsequent studies have shown that radical-aldehyde cyclization to give a six-membered ring can be a highly efficient process, even in the face of a 5-hexenyl alternative [2,3].

SCHEME 1

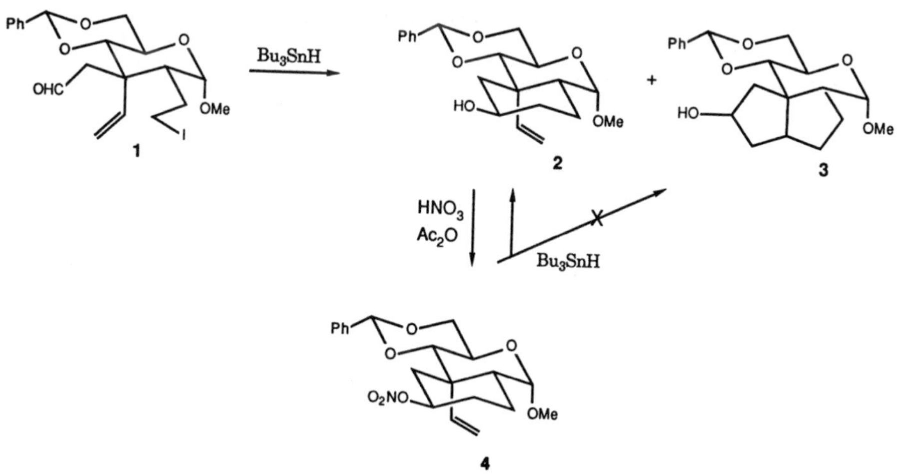

More recent evidence has shown that the formation of cyclohexanols, such as 2, is not reversible under the reaction conditions [4]. Thus, the nitrate ester 4, on reaction with tri-n-butyltin hydride, regenerated 2 cleanly, without any trace of diquinane 3. The latter would have been formed in approximately 20% yield had the oxygen radical from 4 undergone β-scission before closing to regenerate 2.

However, in spite of these early successes, our continuing studies aimed at developing the synthetic potential of the ring closure have met with varied success. The compounds 5, 6, 7, and 8 have been prepared as possible synthetic intermediates for a variety of synthetic targets. In order to obtain a better insight into the relative reactivity of the aldehydo group, vis a vis other radical acceptors, such as nitrile, alkene, and conjugated esters related to (some of) substituents of 5-8 have also been examined.

H. Fischer, H. Heimgartner (Eds.)
Organic Free Radicals
© Springer-Verlag Berlin Heidelberg 1988

SCHEME 2

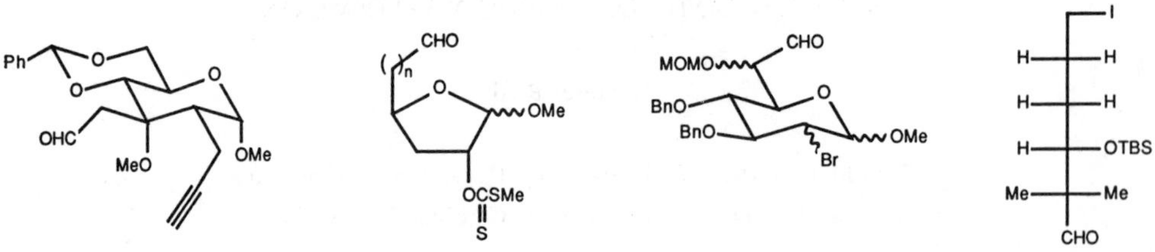

| 5 | 6 | 7 | 8 |

A report on the results obtained with these substrates, as well as related synthetic and mechanistic investigations will be presented.

References:

1   Tsang R, Fraser-Reid B (1986) J Am Chem Soc 108:2116
2   Tsang R, Fraser-Reid B (1986) J Am Chem Soc 108:8102
3   Tsang R, Dickson Jr J K, Pak H, Walton R, Fraser-Reid B (1987) J Am Chem
     Soc 109:3484
4   Vite G D, Fraser-Reid B (1988) Tet Lett 29:1645

RADICAL ADDITION OF ACETONE TO LIMONENE

INITIATED BY KMnO$_4$ IN ACETIC ACID MEDIUM

C. Gardrat

Laboratoire Chimie Appliquée, Université Bordeaux I

Ecole Nationale Supérieure de Chimie et de Physique de Bordeaux

351, cours de la Libération, 33405 Talence Cedex, France.

Abstract : Acetonyl radicals generated by potassium permanganate in acetic acid medium could be added to the exocyclic double bond of limonene. Two main products were obtained, one with a carvomenthene type structure and the second with a limonene one. The former comes from a classical mechanism (addition and H-transfer), the latter from the oxidation of the intermediate radical followed by elimination of a proton.

INTRODUCTION

It has been reported that α-oxoalkyl radicals were obtained from ketones by reaction of potassium permanganate in acetic acid medium (1). We took advantage of this reaction to study the addition of acetonyl radicals to several unsaturated kinds of products such as electron-rich alkenes and cyclenes (2,3). Now, we present our first results dealing with the addition of acetonyl radicals to limonene initiated in similar conditions. These results are compared with those obtained with di-tert-butyl-peroxide (DTBP).

EXPERIMENTAL RESULTS

Acetone, limonene, potassium permanganate (molar ratios 50/1/1) and acetic acid (10% vol.) were placed in a stainless bomb and heated to 120°C during one hour. The white solid was discarded and an ether solution of the products was treated by aqueous sodium carbonate, washed with water, dried and the ether pulled off under vacuum. Analysis of the crude reaction mixture by open-tubular column gas chromatography indicated that a lot of products were formed : nevertheless, two main peaks in the ratio of about 70/30 in order of emergence appeared beside several small ones (<3% each) and a curious large peak. The same qualitative results were obtained when working at room temperature (ten days were necessary to transform all the potassium permanganate). When the quantity of potassium permanganate was doubled, the second main chromatographic peak vanished. Hydrogenation of the reaction mixture over Raney nickel led to the disappearence of the first main peak and augmentation of the second one.

In order to determine whether the structure of the main products would be different with an initiation by a peroxide, we have decided to study the addition of acetone to limonene in the presence of DTBP . The different products (molar ratios : acetone/limonene/DTBP = 10/1/0.2) were heated in a stainless bomb to 150°C during 3 hours.

H. Fischer, H. Heimgartner (Eds.)
Organic Free Radicals
© Springer-Verlag Berlin Heidelberg 1988

The analysis of the reaction mixture by capillary gas chromatography shew the formation of a largely predominant adduct with a retention time identical to the second major peak described above.

The separations of the different main products in the reaction mixtures were performed by chromatography over silicagel.

The chemical reactions and the different spectroscopic data (ir, nmr, ms) were in good agreement with the following structures :

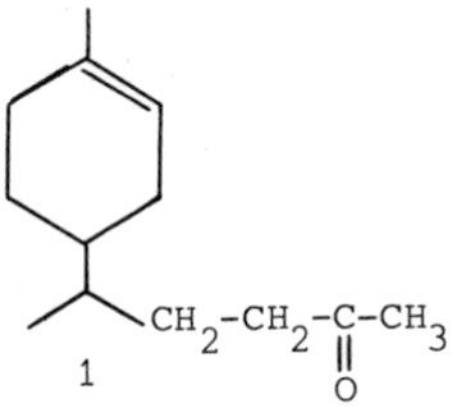

CH_2-CH_2-C-CH_3
O

1

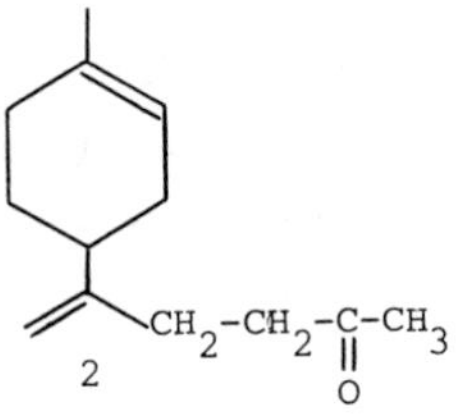

CH_2-CH_2-C-CH_3
O

2

	1	2
$KMnO_4$:	30%	70%
DTBP :	∿100%	traces

SUGGESTED MECHANISM

With DTBP, it seems that the mechanism is quite simple : addition of the acetonyl radical on the exocyclic double bond followed by a transfer of an hydrogen atom coming from the solvent (way a).

With potassium permanganate, there is a competition between two possibilities : a mechanism identical to the previous one (way a) and a second corresponding to the oxidation of the intermediate radical followed by the loss of a proton (way b). The latter predominates when the proportion of the oxidant is increased. Such a mechanism has already been proposed in the case of the addition of acetone to β-pinene (2). It seems that oxidation is favoured with tertiary radicals.

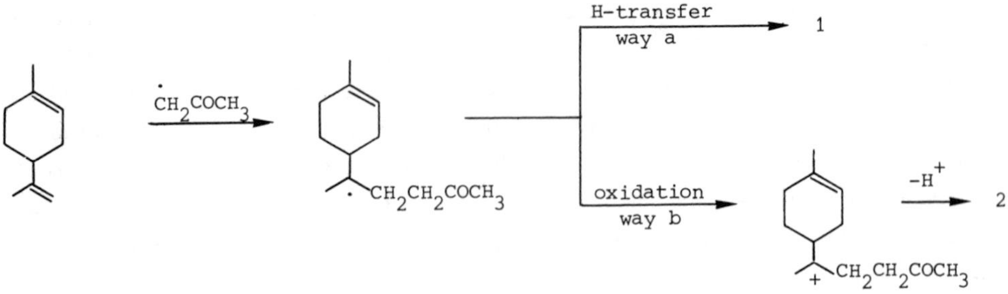

$\dot{C}H_2COCH_3$
$CH_2CH_2COCH_3$
H-transfer
way a
1
oxidation
way b
$-H^+$
2
$CH_2CH_2COCH_3$

References :

1 Vinogradov MG, Direi PA, Nikishin GI (1977) Izv Akad Nauk SSSR, Ser Khim, 7182
2 Gardrat C (1984) 4th Int Symp OFR, Saint Andrews, 36
3 Gardrat C (1986) Euchem Conference OFR, Assise

THREE PARAMAGNETIC REDUCTION STAGES OF PHENYL SUBSTITUTED 1,2:9,10-DIBENZO[2.2]PARACYCLOPHANES

F. Gerson and T. Wellauer

Institut für Physikalische Chemie der Universität Basel,
Klingelbergstrasse 80, CH-4056 Basel, Switzerland

A. de Meijere and O. Reiser

Institut für Organische Chemie der Universität Hamburg,
Martin-Luther-King-Platz 6, D-2000 Hamburg 13, F.R.G.

Abstract: Radical anions, triplet dianions, and radical trianions of phenyl substituted 1,2:9,10-dibenzo[2.2]paracyclophanes **2**, **3**, and **4** have been characterized by ESR, ENDOR, and TRIPLE resonance spectroscopy.

The first synthesis of 1,2:9,10-dibenzo[2.2]paracyclophane (**1**) was reported three years ago [1]. More recently, other workers also prepared **1** by following a different route [2]. In addition, derivatives bearing two (**2** and **3**) or four (**4**) phenyl substituents have become available [2].

1	$R^1 = R^2 = R^3 = R^4 = H$
2	$R^1 = R^2 = Ph, R^3 = R^4 = H$
3	$R^1 = R^3 = Ph, R^2 = R^4 = H$
4	$R^1 = R^2 = R^3 = R^4 = Ph$

Although **1**$^{\cdot-}$ has been studied in detail [3], the formation of multi-charged anions was not detected. The propensity to accept more than one electron is greatly enhanced in the phenyl derivatives **2**, **3**, and **4**, and, accordingly, three paramagnetic reduction stages have been observed with the use of ESR, ENDOR and TRIPLE resonance spectroscopy.

The first reduction step yields the radical anions in which the un-paired electron resides in one of the two "outer" biphenyl (**2**$^{\cdot-}$ and **3**$^{\cdot-}$) or o-terphenyl (**4**$^{\cdot-}$) π-systems orthogonal to the "inner" [2.2]para-cyclophane unit. Depending on the experimental conditions (solvent, counterion, temperature), the electron exchange between the two π–

H. Fischer, H. Heimgartner (Eds.)
Organic Free Radicals
© Springer-Verlag Berlin Heidelberg 1988

systems is slow or fast on the hyperfine time-scale. In the dianions, each of the two outer π-systems bears one negative charge, and the two additional electrons can be paired or unpaired to give a singlet ($\underline{2}^{2-}$, $\underline{3}^{2-}$, and $\underline{4}^{2-}$) or a triplet spin state ($\underline{2}^{\cdot\cdot}$ $\underline{3}^{\cdot\cdot}$, and $\underline{4}^{\cdot\cdot}$) respectively. ESR studies strongly suggest that triplet is the ground state of the dianions with the two unpaired electrons separated by 900-1000 pm. Further reduction leads to the radical trianions in which the third additional electron is accommodated by the inner [2.2]paracyclophane unit.

1 Chan CW, Wong HNC (1985) J Am Chem Soc 107: 4790
2 Reiser O, Reichow S, de Meijere A (1987) Angew Chem 99:1285;
  Angew Chem Int Ed 26:1277; Stöbbe M, Reiser O, Näder R,
  de Meijere A (1987) Chem Ber 120:1667
3 Gerson F, Martin WB Jr, Wong HNC, Chan CW (1987) Helv Chim Acta
  70:79

THE ROLE OF SOLVENT DYNAMICS IN INTER- AND INTRAMOLECULAR ELECTRON EXCHANGE REACTIONS

G. Grampp[*], Institute of Physical Chemistry, University of Erlangen,
D-852 Erlangen, F.R.G.

B. Herold, C. Shohoji, Instituto Superior Tecnico, Lisbon, Portugal

S. Steenken, MPI für Strahlenchemie, Mühlheim, F.R.G.

We report on the solvent dynamic effects of inter- and intra-molecular electron exchange reactions within organic systems:

$$Q^{+/-}_{\bullet} + Q \underset{}{\overset{k}{\rightleftharpoons}} Q + Q^{+/-}_{\bullet} \tag{1}$$

For a nonadiabatic electron transfer (et) the rate constant k is given by the well known expression:

$$k = \frac{2\pi\, V_{12}}{\hbar(4\pi\, \lambda_o RT)^{1/2}} \cdot \exp\left(-\frac{\lambda_i + \lambda_o}{4RT}\right) \tag{2}$$

where V_{12} denotes the resonance energy splitting. λ_i is the inner sphere reorganization energy and λ_o stands for the outer sphere reorganization energy, describing the influence of the solvent:

$$\lambda_o = \frac{z^2 e_o^2\, N_L}{4\pi\, \varepsilon_o} \cdot \left(\frac{1}{r} - \frac{1}{d}\right) \cdot \gamma \tag{3}$$

where r is the molecular radius, d the reaction distance and $\gamma = (1/n^2 - 1/\varepsilon)$ is the solvent parameter. n, ε are the refractive index and the dielectric constant of the solvent.

Therefore for a nonadiabatic et the solvent dependence of k should be given by: $\log k \sim \gamma$. This is fuond for reaction (1) with Q=p-phenylenediamines.

On the other side the solvent dependence of an adiabatic et can be expressed by:

$$k = \tau_L^{-1} \sqrt{\frac{\lambda_o}{16\pi\, RT}} \; \exp\left(-\frac{\lambda_i + \lambda_o}{4RT}\right) \tag{4}$$

where τ_L is the longitudinal dielectric relaxation time of the solvent. Within this limit the solvent dependence is given by: $\log(k\tau_L\gamma^{-1/2}) \sim \gamma$. Such a behaviour was found for Q=TCNE and TCNQ. The decision wheter the et is adiabatic or nonadiabatic depends strongly on the resonance splitting energy V_{12}.

H. Fischer, H. Heimgartner (Eds.)
Organic Free Radicals
© Springer-Verlag Berlin Heidelberg 1988

We show that the different solvent dependences of the et-rates
can be explained by an molecular orbital model, considering the
different transition state structures leading to different
values of V_{12}. Since V_{12} depends exponentially on the reaction
distance d, we decided to measure the intramolecular et rates
within 1,3-disubstituted aromatic radical anions having a
fixed reaction distance d and a fixed V_{12}-value. Since V_{12} is
large for intramolecular et, these reactions follow the normal
TST-theory and showed a solvent dependence, according to
$\log(k\overline{I}) \sim \gamma$. Herein I denotes the moment of inertia of the solvent,
as recently pointed out by Rips and Jortner.

ON THE RADICALIC METHYLATION.

UNEXPECTED FORMATION OF SULFONES AND SULFONAMIDES IN THE REACTION

OF NITROSOARENES WITH FENTON'S REAGENT IN DMSO.

P. Bruni, L. Cardellini, L. Greci, P. Stipa

Dipartimento di Scienze dei Materiali e della Terra,

Facoltà di Ingegneria,

Via Brecce Bianche, I-60131 Ancona, Italy.

It is well known that dimethyl sulfoxide (DMSO) in the presence of the redox system H_2O_2-Fe(II) is one of the most efficient source of methyl radical,[1] and that this method was succesfully used in the methylation of protonated N-heteroaromatic bases.[2] On the other hand it is also well known that alkyl radicals reacting with nitroso derivatives form nitroxides[3] and that these ones, in turn, reacting with alkyl radicals form alkylated hydroxylamines.[4,5] On the basis of the reactions of benzyl and 2-cyano-2-propyl radicals with nitrosobenzenes[6] we planned the transformation of nitroso-derivatives into N,O-dimethylated hydroxylamines by means of the Fenton's reagent.

The reactions, carried out by adding 40% aqueous H_2O_2 to the DMSO solution of nitroso-derivative and $FeSO_4 \cdot 7H_2O$, gave compounds 1a (azoxybenzene), 2a (methansulfonanilide) and 7a (N,O-dimethylphenylhydroxylamine) in the case of nitrosobenzene; whereas compounds 1b (4,4'-dimethylamino-azoxybenzene), 3b [N-(4-dimethylamino-phenyl)-N,N-dimethansulfonamide], 4b [N-(2-methansulfonyl-4-dimethylamino)-phenyl-N-methansulfonamide], 5b [N-(3-methansulfonyl-4-dimethylaminophenyl)-N-methansulfonamide], 6b [N-(2-methensulfonyl-3-methyl-4-dimethylaminophenyl)-N-methansulfonamide] were isolated in the case of the p-nitroso-N,N-dimethylaniline.

It worth noting that the expected N,O-dimethylated hydroxylamine was formed only in the case of nitrosobenzene and the only methylated product from p-nitroso-N,-N-dimethylaniline was 6b. The unexpected results described above indicate that the sulfonylated compounds are the main products with respect to the methylated ones.

H. Fischer, H. Heimgartner (Eds.)
Organic Free Radicals
© Springer-Verlag Berlin Heidelberg 1988

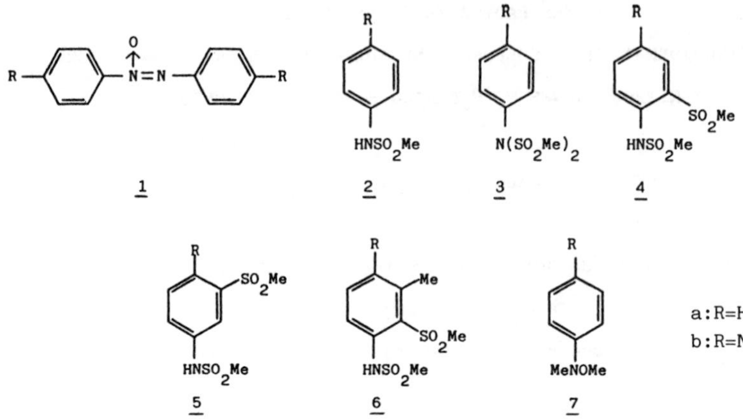

R—⬡—N⁼N—⬡—R (with O↑ above N)

HNSO₂Me

1

⬡ R / HNSO₂Me

2

⬡ R / N(SO₂Me)₂

3

⬡ R / SO₂Me / HNSO₂Me

4

⬡ R / SO₂Me / HNSO₂Me

5

⬡ R / Me / SO₂Me / HNSO₂Me

6

⬡ R / MeNOMe

7

a:R=H
b:R=NMe₂

The formation of sulfonamides and sulfones could be likely explained by reaction of the nitroso compound with sulfinic acid formed by the Fenton's reagent in DMSO as shown in eq.s 1 and 2:[7]

$$H_2O_2 + Fe(II) \longrightarrow HO^- + HO^\bullet + Fe(III) \qquad 1$$

$$HO^\bullet + MeSOMe \longrightarrow Me^\bullet + MeSO_2H \qquad 2$$

Studies are in progress to elucidate the mechanism and the reasons why the N,O-dimethylated hydroxylamine is not formed in the case of the p-nitroso-N,N-dimethylaniline.

References.

1) Giordano C., Minisci F., Fortelli V., Vismara E. (1984) J. Chem. Soc. Perkin Trans. II 293
2) Minisci F., Synthesis (1973) 1; Top. Curr. Chem. 62:1
3) Perkins M.J. (1980) In "Advances in Physical Organic Chemistry"; Gold V., Bethell D., Eds., Academic Press: New York, Vol. 178, p. 1; Gronchi G., Courbls P., Tordo P., Mousset G., Simonet J. (1983) J. Phys. Chem. 87:1343
4) Beckwith A.L.J., Bowry V.W., O'Leary M., Moad G., Rizzardo E., Solomon D.H. (1986) J. Chem. Soc., Chem. Comm. 1003
5) Rozantsev E.G., Sholle V. (1971) Synthesis 190; ibid. (1971) 401; Greci L. (1982) Tetrahedron 38:2435
6) Jackson R.A., Waters W.A. (1960) J. Chem. Soc. 1653; Gingras B.A., Bayley C.H. (1959) Can. J. Chem. 37:988
7) Rudqvist U., Torssell K. (1971) Acta Chem. Scand. 25:2183

THEORETICAL STUDY ON THE STRUCTURE OF β-SUBSTITUTED ETHYL RADICALS

Maurizio Guerra

I.Co.C.E.A., CNR
Via della Chimica 8, Ozzano Emilia (Bo), Italy

The structural parameters of β-substituted ethyl radicals CH_2-CH_2X ($X=CH_3, NH_2, OH, F, SiH_3, PH_2, SH, Cl$) have been determined by accurate ab initio calculations.

At UMP2/DZ+d+BF level of theory, there is no evidence of asymmetrical bridging from the substituent; the arrangement of atoms around C_β being tetrahedral. This result disagrees with the bridged hypothesis proposed by Kochi and Krusic (1) to explain the trend of the hyperfine coupling constants (hfs) when the substituent eclipses the singly occupied 2p orbital. Substituents containing third row elements adopt the eclipsed conformation, having a rotational barrier around $C_\alpha-C_\beta$ of about 2 Kcal/mole. The rotational barrier is less than 1 Kcal/mole for second row elements. For $X=CH_3$ or OH the substituent is rotated by about 60° with respect to the eclipsed conformation. The symmetrically bridged structure is less stable by 40-60 Kcal/mole and 10-20 Kcal/mole for substituents containing second and third row elements, respectively. Thus the rotational motion is favoured as against a shuttle motion (2) of the substituent between the two carbon centers.

The hyperfine coupling constant computed for β-protons is strictly related to the electronegativity of the heteroatom. A third term has been added (3) to the well-known Heller-McConnell relationship (4) to correctly reproduce its angular dependence.

$$hfs(\theta,\alpha) = A + B\cos^2\theta + C\cos\theta\cos\alpha$$

where A and B are not significantly affected by the substituent and C is closely related to the electronegativity of the heteroatom. This relationship explains the observed trend of hfs in β-substituted alkyl

H. Fischer, H. Heimgartner (Eds.)
Organic Free Radicals
© Springer-Verlag Berlin Heidelberg 1988

radicals without invoking any distortion at C_β.

References

1 Krusic PJ, Kochi JK (1974) J Am Chem Soc 96:6715

2 Skell PS, Traynham JG (1984) Acc Chem Res 17:160

3 Guerra M (1987) Chem Phys Lett 139:463

4 Heller C, McConnell HM (1960) J Chem Phys 32:1535

RADICAL REACTIONS
FOR USE IN
ORGANIC SYNTHESIS

Jeff Dener, David J. Hart, Horng-Chih Huang, Frank Seely and Shung Wu

Department of Chemistry, The Ohio State University
120 W. 18th Ave., Columbus, Ohio, 43210, USA

The notion that **intramolecular** free radical carbon-carbon bond-forming reactions might be of use in the synthesis of natural products of modest complexity was introduced by the Julia group in the 1960's.[1] The 1980's have born witness to a rebirth of this idea and free radical cyclizations are now being used regularly in synthesis design.[2,3] Our research in this area has focused on the development α-acylamino radical cyclizations for use in alkaloid synthesis[4] and β-acyloxy radical cyclizations for use in terpenoid synthesis.[5] The first portion of this lecture will describe the development of such reactions for use in total or partial syntheses of heliotridine (**1→2**), gelsemine (**3→4**) and pleurotin (**5→6**).[6] In each case, the radical cyclization accomplishes a carbon-carbon bond construction that would be problematic using classical polar bond-forming reactions.

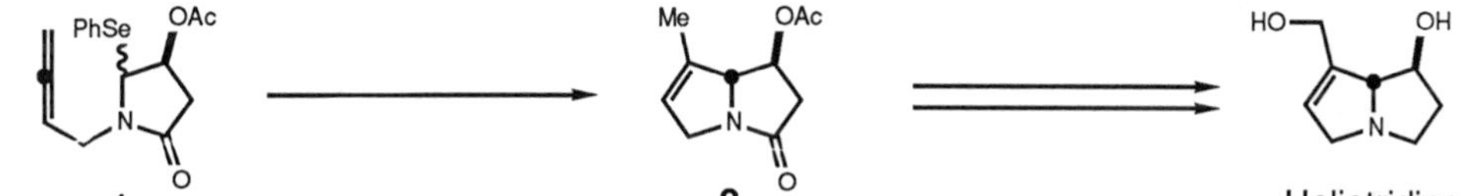

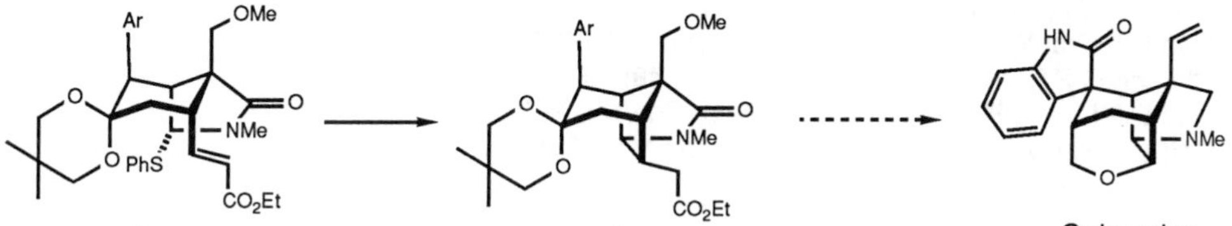

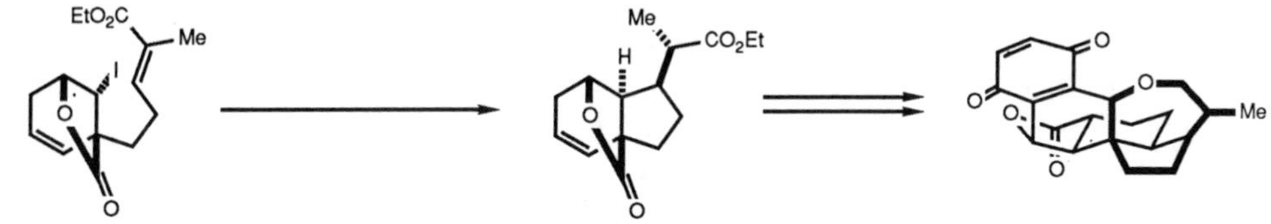

H. Fischer, H. Heimgartner (Eds.)
Organic Free Radicals
© Springer-Verlag Berlin Heidelberg 1988

A number of methods for the construction of carbon-carbon bonds via **intermolecular** free radical addition reactions have also been developed in the past decade.[2,3] Some of these methods require that a large excess of addend be used to obtain high yields of adduct, but several reactions that circumvent this problem have also been developed. These generally involve chain reactions conducted in the absence of good hydrogen atom donors. In this regard, addition-fragmentation and atom (or group) transfer reactions have evolved as "general" strategies for accomplishing near stoichiometric coupling of a variety of radical precursors and addends. We have recently been exploring the use bis(trimethylstannyl)benzopinacolate as a source of trimethylstannyl radicals[7] and have found the this reagent mediates the coupling of halides and selenides with several addends (oxime ethers (**7→8**), α,β-unsaturated esters, α,β-unsaturated nitriles) via what we feel is a free radical process. One useful feature of these reactions is that they proceed in modest to high yield when a 1:1 ratio of free radical precursor is used. Another interesting aspect of this method is that the reactions may proceed via a free radical non-chain mechanism. The second portion of this paper will focus on synthetic and mechanistic aspects of these reactions.

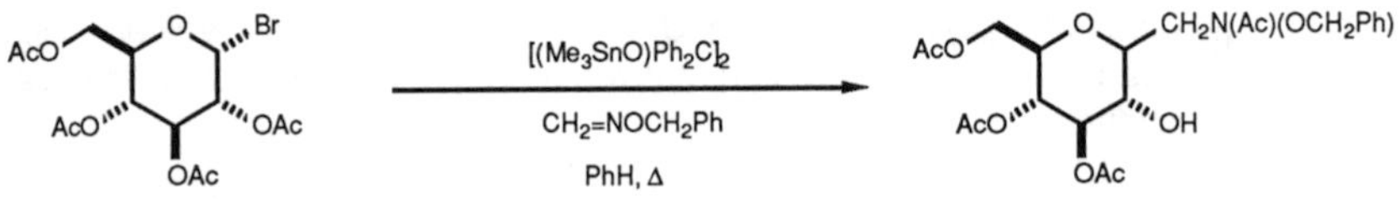

References:

1 Julia M (1964) Rec Chem Prog 25:1
2 Giese B (1986) Radicals in Organic Synthesis: Formation of Carbon-Carbon Bonds, Pergamon Press, New York
3 Ramaiah M (1987) Tetrahedron 43:3541
4 Hart DJ, Tsai Y-M (1982) J Am Chem Soc 104:1430
5 Chuang C-P, Hart DJ (1983) J Org Chem 48:1782
6 Hart DJ, Huang H-C (1988) J Am Chem Soc 110:1634
7 Hart DJ, Seely FL (1988) J Am Chem Soc 110:1631

FREE AND UNFREE RADICAL IONS

Edwin Haselbach

Institut de Chimie Physique
de l'Université, Pérolles,
CH-1700 Fribourg, Suisse.

Electron transfer is a key step in a variety of chemical processes. If the reactants are closed shell neutrals (M), radical
ions ($M^{+\cdot}$ or $M^{-\cdot}$) are initially produced; these may then take
part in various chemical transformations.
Most of these reactions are thermally activated, but a growing
body of evidence has accumulated that radical ions also exhibit
many light induced reactions. These are quite distinct from those
undergone by the neutral parents, being generally initiated by
photons of relatively low energy and often following different
courses.

The present lecture will discuss several topics and questions
concerning $M^{+\cdot}/M^{-\cdot}$ (preparation, detection, identification, structure, energetics, thermal and photochemical reactivity, etc),
using some selected model systems as illustration: "Free" radical
ions prepared in the gas phase or in solution, and "unfree" radical ions prepared and trapped in solid matrices.

H. Fischer, H. Heimgartner (Eds.)
Organic Free Radicals
© Springer-Verlag Berlin Heidelberg 1988

ON THE STABILIZATION OF α-CARBON RADICALS
BY π- AND METHYLTHIO- SUBSTITUENTS

D. Hasselmann and *M. Stepputtis*

Fakultät für Chemie der Ruhr-Universität Bochum
Postfach 102148, D-4630 Bochum, Federal Republic of Germany

The energetic influence of substituents of wide variety on the stability of carbon centered radicals is of profound importance to
mechanistic as well as synthetic organic chemistry. Kinetic and thermodynamic phenomena related to these effects therefore have found
much attention, recently [1]. Whereas the effects of some substituents, e.g. carbomethoxy, cyano, methoxy, and dimethylamino [2] seem
to be established sufficiently those of others are still subject of
controversy [1]. We report on the quantitative effects of substituents falling into this later category, such as the π-substituents
ethynyl, phenyl and *vinyl*, and the *methylthio*-group. α-Radical stabilization energies (α-RSEs) of these substituents have been deduced
from the results of the kinetics of gas phase thermal rearrangements
of 5-substituted *2-methylenebicyclo[2.1.1]hexanes* (**1**), interconver-
sions definitely radical in character.

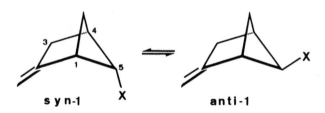

$$X = -C \equiv CH, \ -CH = CH_2, \ -C_6H_5, \ -SCH_3$$

Within the temperature range of 80° to 160°C these substrates undergo
thermal stereomutation, syn-**1** $\rightleftarrows$ anti-**1**, energetically preceding other

H. Fischer, H. Heimgartner (Eds.)
Organic Free Radicals
© Springer-Verlag Berlin Heidelberg 1988

processes of **1**, e.g. fragmentation to X-substituted open chain poly-
enes. Only in the case of the ethenyl substituted compound an about
5 to >10 times slower formation of the bicyclo[3.2.1]-system **3** and
the bicyclo[4.2.1]-system **4**, respectively, competes with the approach
to syn-**1** ⇄ anti-**1** equilibrium.

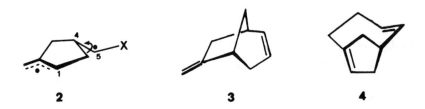

2   **3**   **4**

The epimerizations of **1** will be discussed in terms of participation
of diradicals of type **2** arrived at by homolysis of the C-1-C-5 bond
in **1**. In so far as the bridgehead diene **4** is formed predominantly if
not exclusively from syn-**1** a parallel reaction path for its generation has to be discussed.

Having established the usefulness of the system **1** for the evaluation
of α-RSEs of the substituents mentioned earlier [2] those of the π-
and methylthio substituents investigated now have been deduced from
the enthalpies of activation of the partial reactions syn-**1** → anti-**1**.
Contrary to results from iodination studies on dimethyl sulfide [3]
the methylthio substituent exerts a sizable energetic effect on the
C-1-C-5 bond cleavage in our system. Our α-RSE for X=SCH$_3$ more favourable compares with the results extracted from ESR measurements [4]
and the decomposition of an appropriately substituted azo compound [5].

The kinetic and thermodynamic data found for the X-substituted systems
1 will be analysed in connection with the results obtained from force
field calculations.

1 Viehe HG, Janousek Z, Merényi R (eds) (1986) Substituent Effects in
  Radical Chemistry. Reidel, Dordrecht
2 Hasselmann D, Upadek H, Rottlaender G, Stepputtis M; manuscripts in
  preparation
  Upadek H, Dissertation, Bochum 1977; Rottlaender G, ibid 1984
3 Shum LGS, Benson SW (1985) Int J Chem Kinet 17 277
4 Nonhebel DC, Walton JC (1984) J Chem Soc, Chem Commun 731
5 Luedtke AE, Timberlake JW (1985) J Org Chem 50 268

SEESR OF RADICAL-IONS AND OF
PARAMAGNETIC COMPLEXES

A.V.Il'yasov, M.K.Kadirov, Yu.M.Kargin

Institute of Organic and Physical
Chemistry, Arbuzov str. 8, Kazan,
420083, USSR

Advantages of the simultaneous electrochemical ESR (SEESR) have been used for investigation the mechanisms of electrochemical and subsequent reactions of some organic and phosphorous-organic radical-ions, for determination the structure and kinetic behaviour of the intermediates [1].

High sensitivity of SEESR in the potentiostatic mode, providing the maximum speed of paramagnetic particles generation, allowed us to investigate the series of vinyldiphosphorous radical-anions by means of their ESR spectra exibiting the splittings from ^{13}C nuclei and providing an additional information about the unpaired spin density distribution and the substituents properties.

It has been shown by means of SEESR that fast chemical reaction of dianions of some phenothiazine derivatives took place when the temperature was higher than 300 K. The character of the reaction and the kinetic of the paramagnetic intermediates formation have been established.

The combined electrolysis-ESR method has been applied for the studying the diffusion processes. The potential corresponding to the maximum current has been put on the electrode of the cell and the current flowing through the cell and the ESR signal have been recorded simultaneously. It was found that the character of the ESR signal intensity changes is determined by the diffusion parameters of the system. The equation of the ESR signal intensity for the fast one-electron electrochemical process with linear diffusion to the flat electrode has been obtained: $S \sim Q = FAC\sqrt{Dt/\pi}$, where A- the working electrode surface area, C - concentration of the electrochemically active particles, D - diffusion coefficient, Q - charge, F - the Faraday number.

It appeared that the charge of the electrical double layer and the Faraday currents of by-products have no effect on the ESR signal

H. Fischer, H. Heimgartner (Eds.)
Organic Free Radicals
© Springer-Verlag Berlin Heidelberg 1988

intensity. That allowed us to determine the accurate values of the diffusion coefficients of some organic molecules in an aprotic media.

SEESR application is especially attractive from the point of view of multispin systems as it allows to observe both paramagnetic metal-ion and ligand by varying the ESR signal character and intensity. Complex compounds of Zn(II), Cu(II) and Co(II) with the stable nitroxyl radical 4-(2'-oxopropylidene)-2,2,5,5-tetramethyl-3-imidazolidene-1-oxyl (L) were studied. The spectrum of the biradical system ZnL_2 has been observed at the room temperature. The strong exchange interaction between the paramagnetic centres appeared in the case of the three spin system CuL_2. No ESR spectra were observed for the five spin paramagnetic system CoL_2 at 300 and 77 K due to the very short relaxation times and large fine interactions. The conformational transformations and the sequence of the electron transfer processes for such multispin systems have been established.

Thus, SEESR opens new possibilities for studying the details of the mechanisms of electrochemical processes and the structure of the intermediates.

1 Il'yasov AV, Kadirov MK, Kargin YuM, EichhoffU(1987) Bruker Report 2:20

ESR STUDIES OF 1-SILACYCLOALKYL AND 1-SILA-ALLYL RADICALS

Richard A. Jackson and Antonios Zarkadis

School of Chemistry and Molecular Sciences
University of Sussex, Brighton, BN1 9QJ, U.K.

1-Silacycloalkyl radicals (ring size 4-6), prepared from the corresponding silacycloalkane by γ-irradiation in an adamantane matrix, show a variety of interesting structural features.

1-Silacyclobutyl radicals have a non-planar ring.[1] The unpaired electron on silicon is in a pseudo-equatorial orbital. The most interesting feature of the spectrum is a large long-range coupling of 16.4 G to one of the γ-protons, attributed to the "W-conformation" of the proton in relation to the SOMO on silicon.

1-Silacyclopentyl radicals are believed to have a "twist" rather than an "envelope" conformation, based on the temperature dependence of the esr spectrum which shows an Si-H doublet and two triplets from the β-C-H protons at room temperature, but at low temperatures, all the couplings are doublets. An interesting exchange of hydrogen or deuterium with the adamantane matrix takes place at room temperature.[2]

1-Silacyclohexyl radicals prefer the conformation with the SOMO equatorial, though we have obtained evidence that at low temperatures, there is a small concentration of the axial conformation in equilibrium. The results of these studies, together with previous work, allow conclusions to be drawn about the conformational dependence of C-H couplings β to a radical centre.

Vinylsilane and 1,1-dimethylvinylsilane give rise on γ irradiation to 1-sila-allyl and 1,1-dimethyl-1-sila-allyl respectively. The question of whether these radicals should be regarded as sila-allyl or as vinylsilyl radicals will be discussed.

We thank SERC for a research grant.

References:

1 Jackson RA, Zarkadis AK (1988) J Organomet Chem 341:273
2 Jackson RA, Zarkadis AK (1986) J Chem Soc Chem Commun 205

H. Fischer, H. Heimgartner (Eds.)
Organic Free Radicals
© Springer-Verlag Berlin Heidelberg 1988

PHOTOSENSITIZED GENERATION OF SINGLET OXYGEN ($^1\Delta g$) IN VIVO. PROPOSED BIOMEDICAL APPLICATIONS

G. Jori

Department of Biology, The University
of Padova, 35131 PADOVA, Italy

In the last few years, significant advances in the field of photomedicine (photo-diagnosis and phototherapy) have been obtained by the use of photosensitizers with light-absorption bands in the 600-900 nm interval (e.g. porphyrins, Ps, and phthalo-cyanines, Pcs). The penetrative power of visible light into biological tissues is maximal in this spectral region; the attenuation of incident light to values of 1/e (37%) of the original intensity may often occur at tissue depths larger than 1 cm. Moreover, the normal constituents of tissues, with the exception of melanin, exhibit no appreciable light absorption above 600 nm. As a consequence,only upon adminis-tration of red light absorbing dyes can biological systems be damaged by such wavelengths.

Strategies are being developed for the selective delivery of photosensitizers to predetermined tissues and in particular hyperproliferating tissues, including tumors and atheromas. Thus, some lipoproteins, especially low-density lipopro-teins (LDL), deliver hydrophobic Ps and Pcs to membranous regions of cells having a high mitotic index. Photokinetic and spectroscopic studies show that both Ps and Pcs photosensitize the oxidation of biomolecules in homogeneous solutions through the generation of $O_2(^1\Delta g)$ via energy transfer from the lowest excited triplet state.

In order to define the efficiency of $O_2(^1\Delta g)$-generation in vivo, we performed some studies with Ps and Pcs embedded into model systems, which mimick specific situ-ations occurring during their transport in the bloodstream or interaction with normal and neoplastic tissues. Such systems include:

i) surfactant micelles or small unilamellar liposomes containing the photosensitizer in the hydrophobic regions, while the substrates are located in either
   the apolar compartment or the bulk aqueous medium;

H. Fischer, H. Heimgartner (Eds.)
Organic Free Radicals
© Springer-Verlag Berlin Heidelberg 1988

ii) non-covalent complexes between albumin or LDL and selected Ps and Pcs; in particular, lipoproteins represent ternary systems (water/protein/lipid) with different mutual situations of the substrate and photosensitizer.

The quantum efficiency of $O_2(^1\Delta g)$ generation was estimated by time-resolved luminescence measurements at 1270 nm (corresponding to the decay of $^1O_2$ to ground-state $^3O_2$), as well as by following the photosensitized disappearance of well-known $O_2(^1\Delta g)$-acceptors, such as cholesterol, L-tryptophan and 1,3-diphenyl-isobenzofuran.

The results obtained can be summarized as follows:

i) In all cases the formation of $O_2(^1\Delta g)$ accounts for a major fraction (up to 0.7) of dye triplet-quenching by oxygen.

ii) The efficiency of $O_2(^1\Delta g)$ generation is slightly affected by the microenvironment of the photosensitizer, as long as the latter remains monomeric. The aggregation of Ps and Pcs causes a drop in the yield of $O_2(^1\Delta g)$. The generation of $O_2(^1\Delta g)$ may be also limited by the low local oxygen concentration.

iii) $O_2(^1\Delta g)$ can freely diffuse across protein/lipid or water/lipid interphases. Thus, this excited species can escape from a protein or a lipid matrix and attack oxidizable targets located in the bulk aqueous medium or other cellular compartments.

iv) When the concentration of photooxidizable substrates in the vicinity of the photosensitizer is very large, the overall photoprocess occurs within a small area and the $O_2(^1\Delta g)$ lifetime is dramatically shortened.

1 Jori G (1987) Radiat Phys Chem 30:375
2 Reddi E, Lo Castro G, Biolo R, Jori G (1987) Br J Cancer 56:597
3 Jori G (1987) Photochem Photobiophys, Suppl, 373
4 Lambert C, Reddi E, Spikes JD, Rodgers MAJ, Jori G (1986) Photochem Photobiol 44:595
5 Reddi E, Lambert CR, Jori G, Rodgers MAJ (1987) Photochem Photobiol 45:345
6 Beltramini M, Firey PA, Ricchelli F, Rodgers MAJ, Jori G (1987) Biochemistry 26:6852

CIDNP IN THE REACTIONS OF PHOTOOXIDATION
OF BENZYLIC ACID AND PHENOLS SENSITIZED
BY URANYL ION

A.L.Buchachenko, I.V.Khudyakov, E.S.Klimtchuk, L.A.Margulis

Institute of Chemical Physics, Academy of Sciences of the USSR, 117334 Moscow, USSR

and A.Z.Yankelevich

Karpov Institute of Physical Chemistry, 107120 Moscow, USSR

A search of reversible steps in photochemical reactions of uranyl ion seems to be an important problem of photochemistry and magnetochemistry. The present poster is devoted to the CIDNP study in the reactions of photoexcited uranyl ions with phenols ($\emptyset OH$) and benzylic acid in methanol-d_4. The following OOH have been used: p-cresol, 2,6-di-tert-butylphenol, 2,6-di-tert-butil-4-methylphenol, 2,6-di-tert-butyl-4-ethylphenol.

The solutions have been irradiated by UV-light in a cavity of Bruker HX-90 ER spectrometer (60 MHz) with simultaneous recordering of ^1H and also ^{13}C NMR spectra. It is known that such reactions lead to formation of corresponding aromatic free radicals which may be readily detected by flash photolysis technique /1,2/.

These reactions proceed according to the following mechanism and are accompanied by formation of polarized products:

$$UO_2^{2+} + h\nu \longrightarrow (UO_2^{2+})^* \tag{1}$$
$$(UO_2^{2+})^* + \emptyset OH \longrightarrow (UO_2^+, \emptyset O^\cdot) + H^+ \tag{2}$$
$$(UO_2^+, \emptyset O^\cdot) \xrightarrow{H^+} UO_2^{2+} + \emptyset OH\uparrow \tag{3}$$
$$(UO_2^{2+})^* + Ph_2C(OH)COOH \longrightarrow (UO_2^+, Ph_2\dot{C}OH) + CO_2 + H^+ \tag{4}$$
$$(UO_2^+, Ph_2\dot{C}OH) \xrightarrow{3H^+} U^{4+} + Ph_2CO\uparrow + 2H_2O \tag{5}$$

cf. Fig. 1,2. (The mark $\uparrow$ signifies the polarized product).

Radical pairs (RP) formed according to reactions (2,4) participate in electron transfer reactions (3,5) and escape from a solvent cage. The irradiation of $\emptyset OH$ solutions leads to polarized $\emptyset OH$ but also to polarized labile products of $\emptyset O^\cdot$ dimerization - dimers

H. Fischer, H. Heimgartner (Eds.)
Organic Free Radicals
© Springer-Verlag Berlin Heidelberg 1988

(D). The assignment of signals in ^{1}H NMR of $\varnothing$OH and D has been given elsewhere /3/.

The mechanism of CIDNP formation in the reactions of these four $\varnothing$OH is identical. The general polarization picture is very similar to that in a reaction of $(Ph_2CO)_T^*$ with $\varnothing$OH /4/. We have reproduced data of ref. /4/, where polarization in the system of Ph_2CO + 2,6-di-tert-butylphenol has been studied. This resemblance in the formation of polarization during reactions of $(Ph_2CO)_T^*$ and $(UO_2^{2+})^*$ with $\varnothing$OH enables us to suggest that the reactive state of $(UO_2^{2+})^*$ is a triplet one.

The ^{1}H NMR spectrum of Ph_2CO^+ formed in reaction (5) shows emission of ortho- and para-protons and enhanced absorption of metha-protons (Fig. 2b). It has been found that polarization in the reaction of $(UO_2^{2+})^*$ with benzylic acid is not caused by photochemical reactions of the accumulated in reaction (5) Ph_2CO. Photolysis of Ph_2CO in methanol-d_4 leads to quite different polarization; in particular the Ph_2CO itself stays non-polarized.

The observed signs of CIDNP signals of $\varnothing$OH and Ph_2CO (Fig.1,2) are in agreement with the Kaptein's rule under assumptions that $\varnothing$OH and Ph_2CO are in-cage formed products of the triplet pair and g-factor of organic free radical is higher than that of UO_2^+.

The study of photopolymerization sensitized by $UO_2(NO_3)_2$ has revealed that the moderate external magnetic field accelerates reaction /6/. It has been concluded that magnetic field increases the initiation rate and facilitates a radical escape from a pair $^3(UO_2^+,R^\cdot)$.

Thus we have recorded CIDNP formed in RP $(UO_2^+,R^\cdot)$. The existence of "spin memory" of such pairs is important for the assignment of $(UO_2^{2+})^*$ spin state. There are different point of view: it have been suggested that $(UO_2^{2+})^*$ is in singlet state or it has not a definite spin state due to strong spin-orbit coupling, see e.q. /1,6/. Our data demonstrate that the dominant yield into the reactive state of $(UO_2^{2+})^*$ makes a triplet state which is preserved under a transformation of $(UO_2^{2+})^*$ into corresponding RP.

1 Sergeeva GI, Chibisov AK, Levshin LV, Karyakin AV (1976) J Photochem 5:253
2 Kemp TJ,Martins LJA (1980) J Chem Soc Perkin Trans II : 1708
3 Yankelevich AZ, Khudyakov IV, Buchachenko AL Khim Phisika, in press
4 Schilling MLM (1981) J Amer Chem Soc 103: 3077
6 Jörgensen CK (1979) J Luminescence 18/19 : 63
5 Golubkova NA, Khudyakov IV, Topchiev DA, Buchachenko AL Dokl AN SSSR: in press

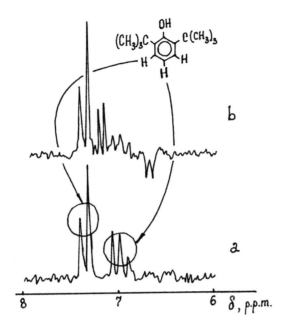

Fig. 1. ^{1}H NMR spectrum of 2,6-di-tert-butylphenol before (a) and
during (b) irradiation

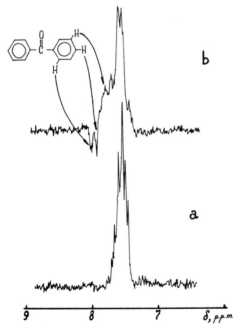

Fig. 2. ^{1}H NMR spectrum of benzylic acid before (a) and during (b)
irradiation

STEREOCHEMICAL ASPECTS OF RADICAL HYDROXYLATION
OF CYCLOHEXENES IN AQUEOUS SOLUTION

G. Koltzenburg

Max-Planck-Institut für Strahlenchemie
Stiftstraße 34-36, D-4330 Mülheim a.d. Ruhr 1, FRG

Abstract: The conformationally biassed cyclohexenes I, II and III were reacted in aqueous solution with $\cdot OH$, $SO_4 \cdot^-$ or $Cl_2 \cdot^-$ radicals to yield directly (with $\cdot OH$) or indirectly, see e.g. (A), four isomeric ß-hydroxycyclohexyl radicals. The latter were converted by H-donors

R—⬡—1 I ;   R—⬡—1 II ;   R—⬡—1 III ;   R = $HO(CH_3)_2C$

(A)   R—⬡ + $Cl_2 \cdot^-$ + H_2O → → $2 Cl^-$ + H^+ + R—⬡—OH

into cyclohexanol derivatives which were analyzed by g.l.c.. The stereochemistry of the entry of OH is compiled in Table 1. The re-

Table 1: Stereochemistry of the entry of OH into the olefinic positions of I, II or III upon radical hydroxylation with $\cdot OH$, $SO_4 \cdot^-$ or $Cl_2 \cdot^-$. Figures (%) in sets of four are arranged according to the pattern:

2-OH(eq) 1-OH(ax)
2-OH(ax) 1-OH(eq)

	I		II		III	
$\cdot OH$	21 29	29 21	29 23	38 10	22 28	28 22
$SO_4 \cdot^-$	4 46	46 4	8 13	77 2	6 44	44 6
$Cl_2 \cdot^-$	4 46	46 4	22 19	32 7	4 46	46 4

sults with $I/SO_4 \cdot^-$ and $I/\cdot OH$ are in the press (1). It may be seen that $\cdot OH$ addition proceeds with distinct but low stereoselectivity favouring axial entry. The indirect hydroxylation proceeds with high stereoselectivity in five of the above six cases. The exemption is $II/Cl_2 \cdot^-$ where the stereoselectivity may be regarded as medium. Indirect radical hydroxylation proceeds through $SO_4 \cdot^-$ or $Cl \cdot$ adduct radicals which hydrolyze through short-lived radical cations as intermediates, (1). To explain the observed stereochemistry it is assumed that adduct radical formation leads first to four radicals, two of which, a and b, possess chair conformation and carry the addend in axial position, the other two, c and d, possess twist-boat

H. Fischer, H. Heimgartner (Eds.)
Organic Free Radicals
© Springer-Verlag Berlin Heidelberg 1988

conformation and carry the addend in pre-equatorial position. The twist-boat conformers of I'-OSO_3^-, I'-Cl and II'-OSO_3^- live at least ten times longer than the not exactly known time for ring inversion of these radicals ($\geqslant 10^{-9}$ s for cyclohexyl at 20° (2)) and may undergo complete conformational relaxation into radicals e and f with chair conformation and equatorial addends. Elimination of SO_4^{--} or Cl^- from the relaxed radicals a, b, e and f leads first to radical cations in relaxed (chair) conformation which add water with high stereoselectivity into axial positions. Elimination of Cl^- or SO_4^{--} from the radicals II'-Cl, III'-Cl and III'-OSO_3^- is faster than conformational relaxation and the non-relaxed radicals among them lead to non-relaxed radical cations in twist-boat conformation. The non-relaxed III'$^+$ is long-lived enough to relax and subsequently add water into axial positions. The non-relaxed II'$^+$ adds water - faster than it can relax - with preference into pre-equatorial positions.

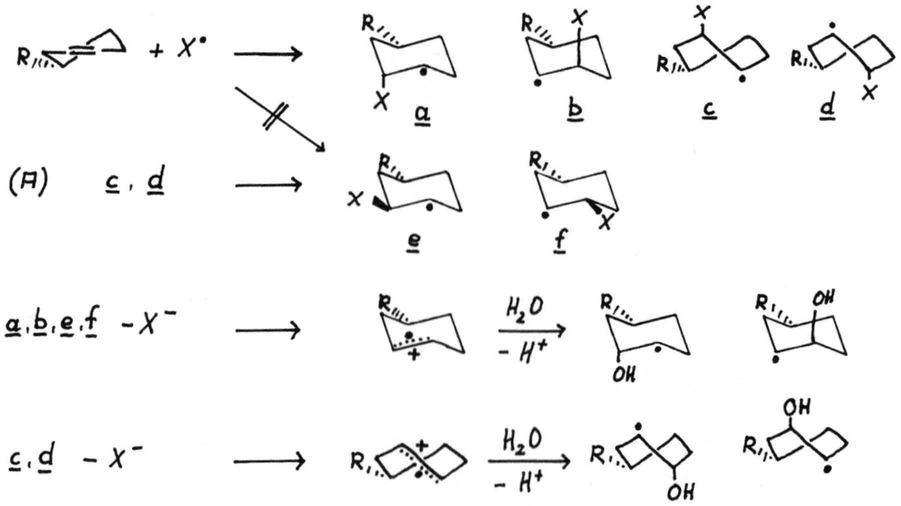

X' = 'OH, SO_4'$^-$, Cl' ; X^- = Cl^-, SO_4^{--}

(CH_3-groups at 1- or 2-position omitted)

One CH_3-group instead of H at the double bond causes an increase of the radical decay rate of 3 or more orders of magnitude: k_{hydrol} for I'-OSO_3^- is $3 \cdot 10^4$ s^{-1}, for II'-OSO_3^- $\geqslant 5 \cdot 10^7$ s^{-1}. Elimination of Cl^- is more than 100 times faster than of SO_4^{--}. The lifetime of I'$^+$ was estimated to 10 - 100 picoseconds (1).

1 Koltzenburg G, Bastian E, Steenken S (1988) Angew Chem in the press
2 Ogawa S, Fessenden RW (1964) J Phys Chem 41:994

REGENERATION OF AMINE
IN CATALYTIC INHIBITION OF OXIDATION

S. Korcek, R. K. Jensen, M. Zinbo, and J. L. Gerlock

Ford Motor Company, Research Staff
P. O. Box 2053, Dearborn, MI 48121 U.S.A.

Abstract: Kinetic and mechanistic investigations of decomposition of O-sec-hexadecyl- and O-3-heptyl-4,4'-dioctyldiphenylhydroxylamines showed that at 120 and 140°C these compounds rapidly decompose to yield 4,4'-dioctyldiphenylamine. These results suggest that in the catalytic inhibition of oxidation by aromatic secondary amines or corresponding nitroxide radicals at elevated temperatures the decomposition of O-sec-alkyldiarylhydroxylamines leads to regeneration of the parent aromatic secondary amine.

It has been previously shown that secondary amine antioxidants and aromatic nitroxide radicals inhibit liquid phase autoxidation of organic substrates by catalytic mechanisms of chain termination. It is generally accepted that their catalytic inhibition cycle involves regeneration of nitroxide radicals.

We have recently investigated the inhibition of oxidation of hexadecane by 4,4'-dioctyl-diphenylamine (Ar_2NH) and by the corresponding nitroxide radical ($Ar_2NO\bullet$) at 160°C. Results of these studies showed that the catalytic cycle is operational even at elevated temperatures. In this cycle $Ar_2NO\bullet$ is formed as an intermediate during the inhibition by Ar_2NH and the parent Ar_2NH is regenerated during the inhibition by both Ar_2NH and $Ar_2NO\bullet$. $Ar_2NO\bullet$ is consumed at a rate proportional to the concentration of isomeric secondary hexadecyl radicals ($R\bullet$).

Based on information available in the literature, the reaction of $Ar_2NO\bullet$ with $R\bullet$ should result in formation of corresponding isomeric O-sec-hexadecylhydroxylamines (Ar_2NOR). Further reactions of O-sec-alkyl-substituted diarylhydroxylamines have not been previously investigated. However, it has been shown that O-tert-alkyl substituted diarylhydroxylamines undergo thermal or peroxy radical induced decomposition leading to formation of $Ar_2NO\bullet$.

In this work we have investigated formation and decomposition of O-sec-alkyl substituted diarylhydroxylamines in more detail. Ar_2NOR have been first synthesized in situ at a lower temperature and then their decomposition have been studied at elevated temperatures.

H. Fischer, H. Heimgartner (Eds.)
Organic Free Radicals
© Springer-Verlag Berlin Heidelberg 1988

Ar$_2$NOR have been prepared from Ar$_2$NO• and secondary alkyl radicals produced from decomposition of t-butylperoctoate (TBPO) in hexadecane at ~90°C. Under these conditions TBPO homolytically decomposes and decarboxylates to generate 3-heptyl radicals (R'•) and t-butoxy radicals (Reaction 1).

$$CH_3CH_2\underset{\underset{CH_3(CH_2)_2CH_2}{|}}{C}HCO_2C(CH_3)_3 \xrightarrow{\Delta} CH_3CH_2\underset{\underset{CH_3(CH_2)_2CH_2}{|}}{C}H\bullet + CO_2 + \bullet OC(CH_3)_3 \tag{1}$$

(TBPO)   (R'•)

t-Butoxy radicals abstract hydrogen from hexadecane to give t-butanol and R• (Reaction 2).

$$\bullet OC(CH_3)_3 + RH \longrightarrow HOC(CH_3)_3 + R\bullet \tag{2}$$

As a result two types of O-sec-alkylhydroxylamines, Ar$_2$NOR and Ar$_2$NOR', are formed from interactions with Ar$_2$NO• (Reactions 3 and 4).

$$R\bullet + Ar_2NO\bullet \longrightarrow Ar_2NOR \tag{3}$$
$$R'\bullet + Ar_2NO\bullet \longrightarrow Ar_2NOR' \tag{4}$$

At elevated temperatures of 120 and 140°C both these O-sec-alkylhydroxylamines were found to decompose and yield mainly Ar$_2$NH and series of isomeric ketones (R$_1$COR$_2$) and minor amount of isomeric alcohols (R$_1$C(OH)R$_2$). In the case of Ar$_2$NOR, formation of these products is consistent with reactions 5 and 6.

$$Ar_2NOR \xrightarrow{k_d} [Ar_2N\bullet + \bullet OR] \begin{cases} Ar_2NH + R_1COR_2 & (5) \\ Ar_2N\bullet + RO\bullet & (6) \end{cases}$$

The k_d values determined from kinetic measurements are 3.4×10^{-4} s^{-1} at 120°C and 3.2×10^{-3} s^{-1} at 140°C. An extrapolation to higher temperatures gives the k_d values of 1.4×10^{-2} s^{-1} for 160°C and 7.0×10^{-2} s^{-1} for 180°C.

Based on these results we conclude that in oxidation systems where secondary alkyl and alkylperoxy radicals are formed, the catalytic inhibition by aromatic secondary amines or corresponding nitroxide radicals at elevated temperatures involves regeneration of the parent amines. This regeneration occurs due to thermal decomposition of intermediate O-sec-alkyl-diarylhydroxylamines.

EPR SPECTRA AND KINETICS OF SOME
CARBONYLOXYL RADICALS, XCO$_2^{\bullet}$

H. Korth, J. Chateauneuf and J. Lusztyk

Division of Chemistry, National Research Council of Canada, Ottawa,
Ontario, Canada, K1A 0R6 and Institute für Organische Chemie,
Universität-GHS Essen, D-4300 Essen, West Germany.

We have demonstrated that aroyloxyl,[1] ArCO$_2^{\bullet}$, and alkoxycarbonyloxyl,[2] ROCO$_2^{\bullet}$, radicals can be generated by laser flash photolysis (LFP) of appropriate peroxides and tert-butyl peresters and that the kinetic behaviour of these radicals can be monitored via a broad absorption they possess in the visible region of the spectrum. The lifetimes of some of these radicals were sufficiently great as to suggest that certain XCO$_2^{\bullet}$ might be observable by conventional EPR spectroscopy during continuous UV irradiation of suitable precursors in solution at low temperatures.

EPR spectra which we assign to CH$_3$CH$_2$CH$_2$OCO$_2^{\bullet}$, Me$_3$CC$\equiv$CCO$_2^{\bullet}$ and trans-Me$_3$CCH=CHCO$_2^{\bullet}$ were obtained by photolysis (1000 W, high pressure H$_g$ lamp) of the corresponding peroxides in Freon 11 or cyclopropane at temperatures <180 K. The first two radicals show just a single line while the last exhibits a doublet of doublets ($\underline{a}^H \approx 2.1$ and 0.4 G). Our assignment of these spectra to the appropriate XCO$_2^{\bullet}$ radical is based on their $\underline{g}$-values (2.0117-2.0128), the resistance of the signals to microwave power saturation and their response to added reactants. Thus, these signals were not eliminated by molecular oxygen but they were quenched by the addition of benzene, toluene, 1,1-diphenylethylene and 1,4-cyclohexadiene, all of which are expected to react rapidly with XCO$_2^{\bullet}$ radicals.[1,2] In addition, Me$_3$CC$\equiv$C^{13}CO$_2^{\bullet}$ has $\underline{a}^{13C}$ = 13.3 G, which indicates a low spin density on the carbon of the carboxyl group, a result which is consistent with the prevailing view that XCO$_2^{\bullet}$ radicals have a σ electronic ground state with the unpaired electron residing mainly on the two oxygen atoms.

These new acetylenic and vinylic carbonyloxyl radicals together with some related species (C$_6$H$_5$CC$\equiv$CO$_2^{\bullet}$, trans-C$_6$H$_5$CH=CHCO$_2^{\bullet}$ and

H. Fischer, H. Heimgartner (Eds.)
Organic Free Radicals
© Springer-Verlag Berlin Heidelberg 1988

$(CH_3)_2C=CHCO_2^{\bullet})$ were shown by 308 nm LFP to have absorptions in the visible similar to those observed previously for $ArCO_2^{\bullet}$ [1] and $ROCO_2^{\bullet}$ [2] radicals. Application of standard kinetic procedures[1] demonstrated that carbonyloxyl radicals generally show increasing reactivities along the series: $C_6H_5CO_2^{\bullet} \sim RCH=CHCO_2^{\bullet} < RC \equiv CCO_2^{\bullet} < ROCO_2^{\bullet}$. We attribute this result to variations in the importance that polar, canonical structures (e.g., $[XCO_2^- \ RH^+]^{\ddagger}$) contribute to the stabilization of the transition state for reaction.

(1) Chateauneuf J, Lusztyk J, Ingold KU (1987) J Am Chem Soc 109:397-899; (1988) 110:2877-2885, 2886-2893
(2) Chateauneuf J, Lusztyk J, Maillard B (in press) J Am Chem Soc

THE STRUCTURE OF Co(II)-CORRINS - 'NATURE'S TRAPS' FOR ORGANIC RADICALS

B.Kräutler[1], W.Keller[2] & C.Kratky[2]

[1]Laboratorium f.Organische Chemie der ETH, ETH-Zürich,
Universitätstrasse 16, CH-8092 Zürich, Switzerland
and
[2]Institut f. Physikalische Chemie, Universität Graz,
Heinrichstrasse 28, A-8010 Graz, Austria

The biosynthetic functions of coenzyme B_{12} (**1**, adenosyl-cobalamin) are based on its ability to act as a "free radical reservoir" [1], and on the complementary reactivity of cob(II)-alamin (**2**,'B_{12r}') as well, to function as a stable and efficient trap for organic radicals in nature. Co(II)corrins, such as **2**, are stable radicaloid species in oxygen-free homogeneous solution and combine with organic radicals at nearly diffusion controlled rates to give organo-Co(III)corrins, such as **1** [2].

Cob(II)alamin (**2**) was prepared by catalytic hydrogenation of aquocobalamin and was crystallized from deoxygenated aqueous acetone. X-ray analysis of anaerobically sealed crystals of **2** furnished the three dimensional molecular structure displayed in Figure 1: Accordingly the Co(II)corrin **2** contains a pentacoordinate Co(II)center, to which the nucleotide base binds axially.

As analyzed similarly earlier, but with a nucleotide free Co(II)corrin [3], the transition from the Co(II)corrin **2** to the organo-Co(III)corrin **1** (which would correspond to the trapping of the adenosyl-radical) is accompanied by minor changes only in the three dimensional structures of the corrin ligand and of the nucleotide base and on their arrangement around the central

H. Fischer, H. Heimgartner (Eds.)
Organic Free Radicals
© Springer-Verlag Berlin Heidelberg 1988

cobalt-ion. This structural invariance presumably is a significant factor in support of the low activation barrier observed in Co(II)corrin - radical recombination reactions.

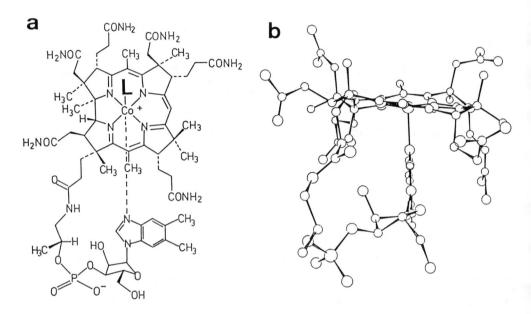

Figure 1: a.Structural formulae of coenzyme B$_{12}$ (**1**, Co-L = Co(III)-5'-adenosyl) and of cob(II)alamin (**2**, Co-L = Co(II))

b.Structure of cob(II)alamin (**2**) as determined by x-ray analysis

References

[1] J.Halpern (1985) Science **227**:869

[2] J.F.Endicott & T.L.Netzel (1979) J.Am.Chem.Soc. **101**:4000

[3] B.Kräutler, W.Keller, M.Hughes, C.Caderas & C.Kratky (1987) J.Chem.Soc.,Chem.Commun. 1678.

ASPECTS OF ORGANOMETALLIC FREE RADICAL CHEMISTRY

M.F. Lappert

School of Chemistry and Molecular Sciences,
University of Sussex, Brighton BN1 9QJ, U.K.

Our group is primarily concerned with synthetic and structural organometallic chemistry. In some systems, paramagnetic organometallic compounds are made and characterised. In some others, paramagnetic transients are implicated in organometallic reactions. Examples of both will be considered.

The methods that we have used for these two types of experiments (apart from synthesis, NMR, and X-ray) involve EPR, cyclic voltammetry (CV), controlled potential electrolysis (CPE) with *in situ* EPR, or stereochemical/kinetic data.

A class of compound which we have used in some of our investigations is an electron-rich olefin, such as (1)-(3). These are not only powerful reducing agents with a first ionisation potential of *ca.* 6 eV, but also are a source of transition metal (M) complexes of type (4). CV Experiments for a series of electron-rich olefins reveal that in some cases there is a 2-electron reduction [*e.g.*, to yield $(1)^{2+}$], whereas in other cases the one-electron intermediate [*e.g.*, $(1)^{+}$] can also be identified. Data will be presented showing $E_{1/2}$ values for these reductions as a function of structure [1]. In those cases where the cation radicals $(1)^{+}$ were detected, a complementary series of CPE/EPR experiments has made it possible to record their EPR spectra and to identify all the possible coupling constants involving ^{1}H and ^{14}N.

A phosphorus analogue (5) of a compound of type (1) has been made [2].

Data will be presented on the $M^{IV} \rightarrow M^{III}$ reduction for a series of complexes of formula $[MCp_2X_2]$ (M = Ti, Zr, or Hf; $\bar{C}p$ = a η-cyclopentadienyl-type ligand; X = F, Cl, Br, or I), and also of some related compounds containing alkyl or phosphorus ligands [3]. It should be noted that well characterised Zr^{III} compounds are exceedingly rare.

Spectroscopic and EPR results will be presented on some novel organometallic Th^{III} complexes (*cf.*, ref. 4).

H. Fischer, H. Heimgartner (Eds.)
Organic Free Radicals
© Springer-Verlag Berlin Heidelberg 1988

Experiments will be described relating to the reduction of $H_2[PtCl_6]xH_2O$ by the disiloxane $[Si(CH=CH_2)Me_2]_2O$, leading to a Pt^0 complex (*cf.*, ref. 5). Evidence will be presented relating to possible intermediates in the reduction [6]. The results are relevant to the important Pt-catalysed hydrosilylation of alkenes.

It is hoped to report some new results on persistent main group metal-centred radicals, including $\dot{A}sR_2$ [R = $CH(SiMe_3)_2$] (*cf.*, ref. 7 for an early paper).

References

1 Campbell GK, Skyropoulos K Unpublished work
2 Anderson DM, Hitchock PB Unpublished work
3 Antiñolo A, Campbell GK, Winterborn DMW Unpublished work
4 Kot WK, Shalimoff GV, Edelstein NM, Edelman MA, Lappert MF (1988) J Am Chem Soc 110:986
5 Chandra G, Lo PB, Hitchcock PB, Lappert MF (1987) Organometallics 6:191
6 Hitchcock PB, Warhurst NJW Unpublished work
7 Gynane MJS, Hudson A, Lappert MF, Power PP, Goldwhite H (1980) J Chem Soc Dalton Trans 2428

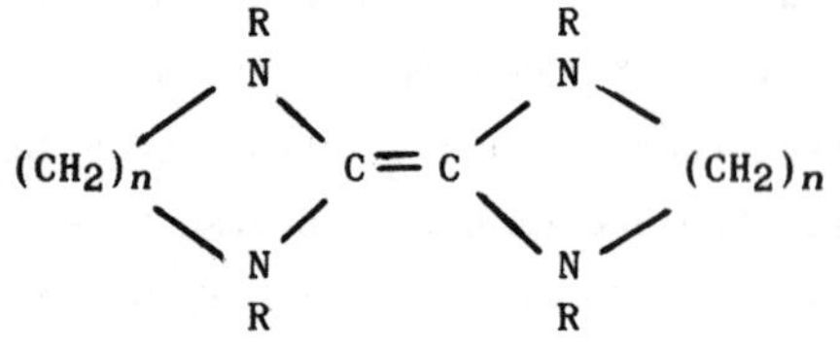

(1)

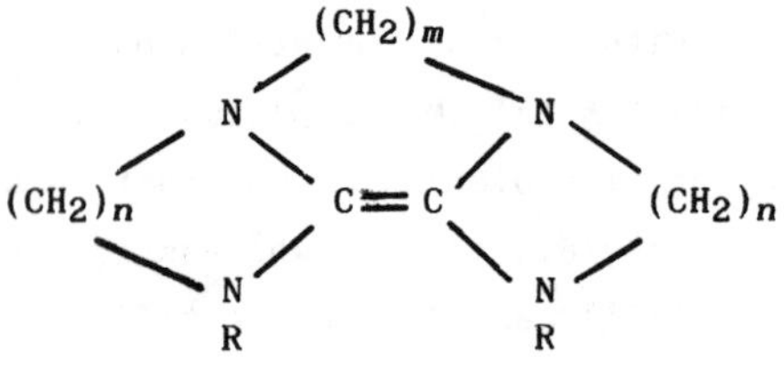

(2) (*n* = 2 or 3)
(*m* = 2 or 3)

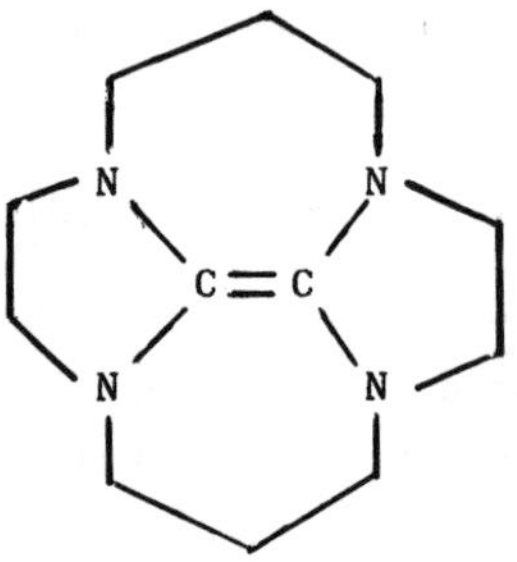

(3)

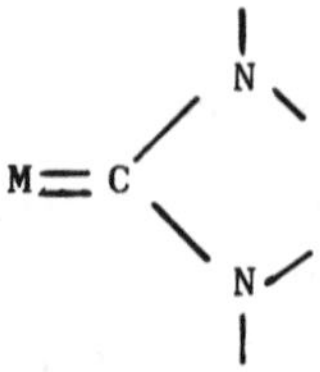

(4)

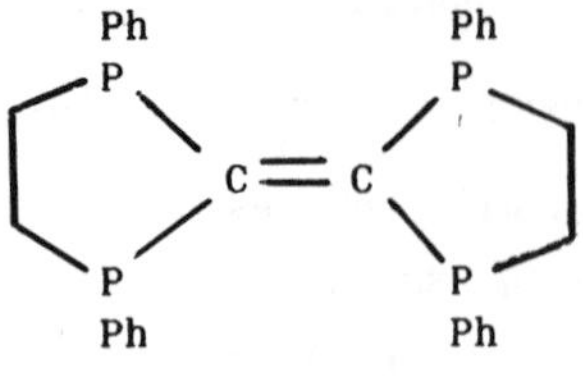

(5)

ENDOR INVESTIGATIONS ON THE SUBSTITUENT DEPENDENCE OF THE STABILISATION OF TRIPHENYLMETHYL RADICALS

M.Lehnig and U.Stewen

Universität Dortmund, Germany

Abstract: The influence of polar substituents on the spin density in $\underline{1}$ and $\underline{2}$ is less or equal to that of the phenyl group. Radicals $\underline{2}$ with "push-pull" substituents show a small capto-dative effect.

The stabilisation of free radicals by polar substituents as well as the capto-dative stabilisation by combination of electron accepting and donating substituents is discussed intensively and controversially during the last years. Spin density distributions in radicals $\underline{1}$ and $\underline{2}$ should react on substituents. As the complex ESR spectra could not be analysed in detail (1), ENDOR investigations have been performed.

$\underline{1}:$ [structure] $\underline{2}:$ [structure]   R , R´ =   t-Bu-,-OMe,-OPh   -CF$_3$, -CN,-COPh

Table $\underline{1}$ shows the ENDOR data of radicals $\underline{1}$, dissociation enthalpies and degrees of dissociation of the dimers (1) are added. From the

Table 1. ENDOR data of radicals $\underline{1}$ in toluene at 200 K

R	a_p (G)	a_o (G)	a_m (G)	a_R (G)	H_{diss} [a]	α [b]
OMe	2.93	2.58 [c]	1.02/1.16	H: 0.31		0.24
H [d]	2.86	2.61	1.14		10.7 ± 0.2	0.12
t-Bu	2.85	2.60	1.14	H: 0.11	10.2 ± 0.3	0.18
OPh	2.84	2.60 [c]	1.12 [c]	H: 0.05		0.16
CF$_3$	2.76	2.54 [c]	1.13 [c]	F: 4.68	10.5 ± 0.2	0.17
Ph [d]	2.72	2.48/2.72	1.10/1.21	H: 0.19/0.49		0.31
CN	2.62	2.38/2.86	1.06/1.16	N: 0.47 [e]	10.0 ± 0.2	0.28
COPh	2.60	2.41/2.60	1.08/1.23	H: < 0.02	10.2 ± 0.2	0.33

[a] Dissociation enthalpy of the dimer in kcal/mol (1). [b] Degree of dissociation of the dimer at 298 K, 0.01 M in benzene (1). [c] Further splittings not resolved. [d] Taken from (2). [e] ESR data.

H. Fischer, H. Heimgartner (Eds.)
Organic Free Radicals
© Springer-Verlag Berlin Heidelberg 1988

Table 2. ENDOR data of radicals $\underline{2}$ in toluene at 200 K

R/R′	a_p (G)	a_o (G)	a_m (G)	a_R (G)	$a_{p,calc} - a_p$
OMe/OMe	2.92	2.57 [c]	1.04 [c]	H: 0.32	0.08
t-Bu/t-Bu	2.88	2.59	1.13	H: 0.10	-0.04
OPh/OPh	2.83	2.62 [c]	1.10	H: 0.05	-0.01
CF_3/CF_3	2.70	2.53 [c]	1.13	F: 4.36	-0.04
Ph/Ph [d]	2.60	2.38/2.60	1.07/1.17	H: 0.19/0.46	-0.01
CN/CN	2.64	2.30/2.64	1.12	N: 0.42 [e]	-0.26
COPh/COPh	2.46	2.28/2.64	1.04/1.16	H: < 0.02	-0.10
t-Bu/CF_3	2.73	2.52	1.13 [c]	F: 4.73 H: 0.09	0.02 (0.06)
t-Bu/CN	2.60	2.41/2.88	1.07/1.21	H: 0.09	0 (0.16)
OMe/CN	2.53	2.35/2.85	0.96/1.20	H: 0.32	0.15 (0.25)

a_p values it is concluded that t-Bu, OMe, and OPh have no or only a small influence on the spin density distribution in the unsubstituted phenyl groups of $\underline{1}$ ($\lesssim$ 2.5 %), CN and COPh diminish the spin density by 9 %, CF_3 by 3.5 %. The dissociation enthalpies are not influenced by the substituents. The degrees of dissociation, however, reflect the spin withdrawing effect of the substituents, an exception is the OMe group. This might be a consequence of the diminished spin densities in α- and p-position which reduces the rate of recombination.

In Table 2, ENDOR data of radicals $\underline{2}$ are given as well as differences between expected and observed a_p values calculated with Fischer's formula (3) and the a_p values from Table 1. The effect of equal substutuents is less than additive. The two combinations t-Bu/CF_3 and t-Bu/CN show additivity of the substituent effects, the combination OMe/CN leads to a greater than additive reduction of the spin density in the unsubstituted ring. The substituent effect of the first two combinations might be interpreted as being capto-dative, too, if half the a_p values of $\underline{2}$ (R = R′) are used in Fischer's formula (values in brackets). The dissociation enthalpies of the dimers (6.8 - 8.2kcal/mol) (1) are not correlated with the a_p values.

(1) W.P.Neumann,W.Uzick,A.K.Zarkadis, J.Am.Chem.Soc.108(1986)3762.
(2) A.H.Maki et al.,J.Am.Chem.Soc.90(1968)4225.
(3) H.Fischer, Z.Naturforsch. 19a(1964)866, 20a(1965)428.

HYDROGEN ATOM TRANSFER BETWEEN
KETYL RADICALS AND KETONES

John S. Lomas

ITODYS - University of Paris 7 (UA34)
1 rue Guy de la Brosse, 75005 PARIS, France

Abstract: A fraction of the ketyl radicals formed in the thermolysis of tri-t-alkylmethanols transfer hydrogen to aryl ketones while the remainder apparently react with t-alkyl radical inside the solvent cage.

Triplet states of ketones abstract hydrogen from hydrocarbons (SH) to give ketyl and hydrocarbon radicals which then react with formation of pinacols, hydrocarbon dimers and cross-coupling products (1,2) :

$$R_2C=O \ + \ SH \ \xrightarrow{h\nu} \ S_2 \ + \ R_2SCOH \ + \ (R_2(OH)C-)_2$$

Ketyl and t-alkyl radicals generated simultaneously by rate-determining C-C bond fission in the thermolysis of tri-t-alkylmethanols abstract hydrogen from aromatic hydrocarbons to give ketones, secondary alcohols, solvent dimers and analogous cross-coupling products, but no pinacols (3÷5). Provided all R are bridgehead alkyl groups it is possible to account for the product composition on the basis of the following scheme :

$$R_3COH \ \longrightarrow \ R_2C^{\bullet}OH \ + \ R^{\bullet}$$

$$R^{\bullet} \ + \ SH \ \longrightarrow \ RH \ + \ S^{\bullet}$$

$$R_2C^{\bullet}OH \ + \ SH \ \longrightarrow \ R_2CHOH \ + \ S^{\bullet}$$

$$R_2C^{\bullet}OH \ + \ S^{\bullet} \ \longrightarrow \ R_2C=O \ + \ SH$$

$$R_2C^{\bullet}OH \ + \ S^{\bullet} \ \longrightarrow \ R_2SCOH$$

$$R_2C^{\bullet}OH \ + \ R^{\bullet} \ \longrightarrow \ R_2C=O \ + \ RH$$

$$2 \ S^{\bullet} \ \longrightarrow \ S_2$$

$$2 \ R_2C^{\bullet}OH \ \longrightarrow \ R_2C=O \ + \ R_2CHOH$$

Two questions arise, however : (i) In cases where a product can in principle be obtained by more than one route which is/are in fact operative ? (ii) To what extent and in what manner does the initial radical pair reaction before escaping from the solvent cage ?

H. Fischer, H. Heimgartner (Eds.)
Organic Free Radicals
© Springer-Verlag Berlin Heidelberg 1988

An attempt has been made to provide some answers by studying the thermolysis of Ad_3COH, **1**, Ad_2t-BuCOH, **2** (Ad = 1-adamantyl) and other alcohols in toluene-h_8 and d_8 in the presence of aromatic ketones, notably benzophenone. Alcohols **1** and **2** are thermolysed a similar rates at 165°C with the former giving exclusively $Ad^.$ and the latter essentially t-Bu$^.$. The major difference in their products is the higher yield of Ad_2CHOH from **2**, resulting probably in part from H-atom abstraction by the ketyl radical from t-Bu$^.$:

$$Ad_2C^.OH + t\text{-}Bu^. \longrightarrow Ad_2CHOH + C_4H_8$$

Yields of Ad_2CHOH and S_2 are in both cases lower in toluene-d_8 than in normal toluene, while the small isotope effect on the cross-product is in the opposite direction. Progressive addition of benzophenone, BP, to the reaction mixture diminishes the contribution of processes involving H-atom transfer or addition to $Ad_2C^.OH$ with, in the case of **1**, complete elimination of Ad_2CHOH and Ad_2SCOH for mole ratios greater than about 0.4 (BP/**1**). Below this limit Ph_2SCOH is obtained in 75-80% yield (relative to BP) rising to a maximum of 31%, relative to **1**, and that of S_2 decreases from 30 to 7%. These results imply hydrogen abstraction from the initial ketyl radical by benzophenone, followed by reaction of the benzophenone ketyl radical with solvent radical :

$$BP + R_2C^.OH \longrightarrow BPH^. + R_2C=0$$

$$BPH^. + S^. \longrightarrow Ph_2SCOH$$

Similar phenomena are observed for **2** except that the Ph_2SCOH yield reaches its ceiling (39%) at a slightly higher mole ratio (0.5) and that Ad_2CHOH bottoms out to a stationary value of 8% both in normal and deuterated solvent, suggesting a cage effect, i.e. H-atom transfer from one radical to another inside the solvent cage. However, the fact that yields are insensitive to large BP excess suggests that the cage effect is substantially greater and may even account for all the ketone formed in the absence of BP, as well as for this threshold Ad_2CHOH value. Similar minima have been found for other alcohols, tending to decrease slightly as the temperature increases.

For alcohols containing 1-norbornyl groups, where ring-opening occurs (4), this too is inhibited by the addition of benzophenone, indicating that it occurs outside the cage.

1 Hammond GS, Baker WP, Moore WM (1961) J Am Chem Soc 83:2795
2 Wagner PJ, Truman RJ, Puchalski AE, Wake R (1986) J Am Chem Soc 108:7727
3 Lomas JS, Dubois JE (1982) J Org Chem 47:4505
4 Lomas JS (1985) J Org Chem 50;4291
5 Lomas JS (1988) Acc Chem Res 21:73

FORMATION OF DIBENZODIOXINS AND
DIBENZOFURANS IN HOMOGENOUS
GAS-PHASE REACTIONS OF PHENOLS

J.G.P. Born, R. Louw and P. Mulder

Center for Chemistry and the Environment, Gorlaeus Laboratories,
Leiden University, P.O. Box 9502, 2300 RA Leiden, the Netherlands

The operation of Municipal Waste Incinerators (MWI's) results in the emission of organochlorine compounds including trace amounts of hazardous polychlorinated di-benzodioxins and dibenzofurans. Although the presence of PCDDs and PCDFs in stack gases and on fly ash is well established, little is known about the mechanisms and kinetics of formation and the phase(s) - e.g. pyrolysis, burning, or fly-ash cata-lysis - in which these compounds and/or their precursors are formed. Proper insight into these chemical features may learn how to improve a MWI installation so as to reduce, or eliminate, these emissions.

We are now investigating the formation of dibenzofurans (DFs) and dibenzodioxins (DDs) from phenols in the homogenous gas phase under slow combustion and pyrolytic conditions between 500-700°C using flow reactor systems. With a mixture of phenol and $\underline{o}$-, $\underline{m}$- and $\underline{p}$-chlorophenol (CP) DF, four MCDF and ten DCDF were observed as important arene derived products. Identification by GC/MS of the DCDFs formed on reacting specific CP isomers has revealed that at the conditions employed dimeri-zation of phenoxy radicals is the most likely pathway leading to chlorinated DFs. Carbon-carbon bond formation is considered as the first step followed by subsequent rearrangement of the intermediate product and further conversion as depicted below. The chlorine substitution pattern in the DF molecule produced, is governed by the

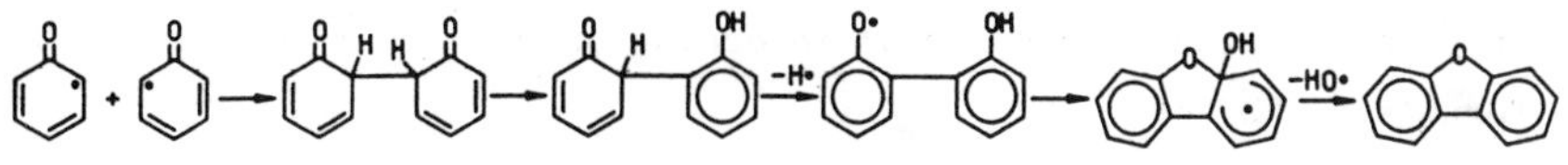

initial carbon-carbon bond formation. A thermochemical-kinetic estimation using group additivity contribution derived from comparison of similar compounds with known heats of formation substantiates this interpretation (see figure). The overall biomolecular rate constant for DF formation from two phenoxy radicals (k_{DF}) is derived from phenol autoxidation data. Assuming a non-chain degradation of the starting compound our kinetic model discloses that at 677°C and $[C_6H_5O\cdot]=10^{-7}M$, k_{DF} becomes ca. 10^6 $M^{-1}.s^{-1}$. Interestingly $\underline{o},\underline{o}'$-dihydroxy biphenyl is quantitatively converted into DF upon pyrolysis at ca. 500°C.

H. Fischer, H. Heimgartner (Eds.)
Organic Free Radicals
© Springer-Verlag Berlin Heidelberg 1988

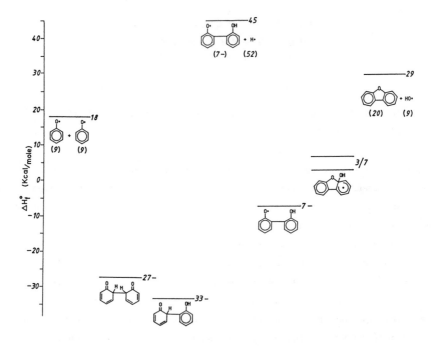

Figure: Energy diagram for dibenzofuran formation from two phenoxy radicals

Experiments involving o-CP showed additional formation of dioxins. Pyrolysis of o-CP per se yields a DD:MCDD:MCDF:DCDF ratio of about 47:0:2:15, whereas under oxidative conditions the product ratio becomes DD:MCDD:MCDF:DCDF = 7:5:8:80. Dioxin formation is thought to begin by ortho carbon-oxygen coupling of two phenoxy radicals; ring closure to produce the dibenzodioxin skeleton appears to be possible only when the reacting phenoxy radicals contain a removable (ortho) substituent such as chlorine. Presence of molecular oxygen in the reaction mixture favours formation DCDF over DD.

FRAGMENTATION PATTERNS FOR ANION RADI-
CALS IN CARBOXYLIC ACIDS

A. Lund

Department of Physics, IFM, University of
Linköping S-581 83 Linköping, Sweden

Abstract: The thermal fragmentation of anion radicals present after X-irradiation of crystalline acetic acid and acetic acid-d_4 has been investigated by means of ESR. The following reaction has been observed:
$$CD_3COOD^- \longrightarrow CD_3\dot{C}O + OD^-$$
Previous results show that carboxylic acid anion radicals containing additional functional groups (-OH, -Cl, -Br, NH_2-) decompose by elimination of water, halogen ion and ammonia, respectively.

INTRODUCTION

Anion radicals are formed as primary reduction products by irradiation of carboxylic acids (1). Anion radicals are reactive, but they can be trapped in the solid state at low temperature. The decomposition to more stable radicals is of interest from a mechanistic point of view. Anion radicals of amino acids, halogenated acids and hydroxy acids decompose by deamination (1), dehalogenation and elimination of water (2), respectively. In this report a previous suggestion (3) that the acetic acid anion radical decomposes by loss of hydroxyl ion is confirmed by observing the formation of the acetyl radical with ESR.

RESULTS AND DISCUSSION

CD_3COOD single crystals grown in quartz tubes were irradiated with X-rays at 77^3K. The ESR spectrum of CD_3COOD^- disappeared after warming to 173 K. The spectrum in Fig. 1 was obtained after warming. It contains two components. One is due to $\dot{C}D_2COOD$. The singlet is assigned to $CD_3\dot{C}O$ based on the following arguments. 1) The g factor is anisotropic. The principal values $g_x = 1.9954$, $g_y = 1.9968$, $g_z = 2.0018$ are in accordance with the values found for acetyl in other systems (4). 2) At high resolution the singlet peak splits into 7 components separated by 1 G as expected for the hyperfine interaction with the D atoms in the CD_3 group. 3) Mechanistic aspects support the identification (3).

$$CD_3COOD \xrightarrow{X} CD_3CO\dot{\,} + D^+ + e^- \qquad (1)$$
$$CD_3COO \rightarrow \dot{C}D_3 + CO_2 \qquad (2)$$
$$\dot{C}D_3 + CD_3COOD \rightarrow CD_2COOD + CD_4 \qquad (3)$$
$$CD_3COOD + e^- \rightarrow CD_3COOD^- \qquad (4)$$
$$CD_3COOD^- \rightarrow CD_3\dot{C}O + OD^- \qquad (5)$$

The $CD_3CO\dot{\,}$ radical in (1) is a (hypothetical) oxidation product which decomposes according to (2) and (3). Thus, $\dot{C}D_2COOD$ and $CD_3\dot{C}O$ are secondary oxidation and reduction products. Inspection of Fig. 1 suggests that the two radicals are present in equal amounts as expected. The results support the proposal that radical ions of unsubstituted carboxylic acids decompose by elimination of hydroxyl ion (5).

H. Fischer, H. Heimgartner (Eds.)
Organic Free Radicals
© Springer-Verlag Berlin Heidelberg 1988

1 Box HC (1977) Radiation Effects. ESR and ENDOR analysis. Academic Press, New York
2 Lund A, Nilsson G, Samskog PO (1986) Radiat Phys Chem 27:111
3 Ayscough PB, Mach P, Oversby JP, Roy AK (1971) Trans Faraday Soc 67:360
4 Landolt-Börnstein Zahlenwerte und Funktionen, Neue Serie Band 9B
  Fischer H, Hellwege KH (eds) (1977) Springer, Berlin, Heidelberg, New York
5 McCalley PC, Kwiram AL (1970) a) J Chem Phys 53:2541
  b) J Am Chem Soc 92:1441

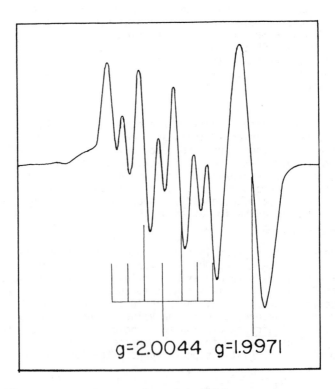

g=2.0044 g=1.9971

Fig. 1   ESR spectrum of X-irradiated CD_3COOD crystal
with lines due to $\dot{C}D_2COOD$ (g=2.0044) and $CD_3\dot{C}O$ (g=1.9971)

UNUSUAL TRANSANNULAR RADICAL CYCLISATIONS

Finlay MacCorquodale and John C. Walton

Department of Chemistry, University of St. Andrews,
St. Andrews, Fife, KY16 9ST, Scotland

Abstract: Cyclohept-4-enylmethyl radical rearranges to bicyclo[3.2.1]
-octyl radical; cyclo-oct-4-enylmethyl radical rearranges to bicyclo-
[4.2.1]- and bicyclo[3.3.1]nonyl radicals. These rearrangements offer
a new synthetic route to bicyclo[3.2.1]octanes and bicyclo[4.2.1]-
nonanes.

Transannular cyclisation of cycloalkenyl alkyl radicals can occur if
there exists a highly populated conformation in which the radical centre
can approach above the plane of the double bond. Normally this requires
that the alkyl chain be more than one carbon atom long.[1] We have
observed two such radicals with a C1 side chain which rearrange to give
bicyclic products.[2]

Reduction of cyclohept-4-enylmethyl bromide with $^{n}Bu_3SnH$ gives products (3) and (4). Radicals (1) and (2) have been observed by ESR.

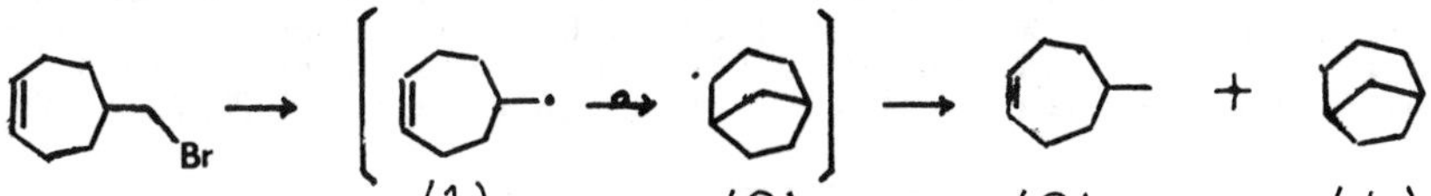

Likewise, reduction of cyclo-oct-4-enylmethyl bromide gives (8), (9)
and (10).

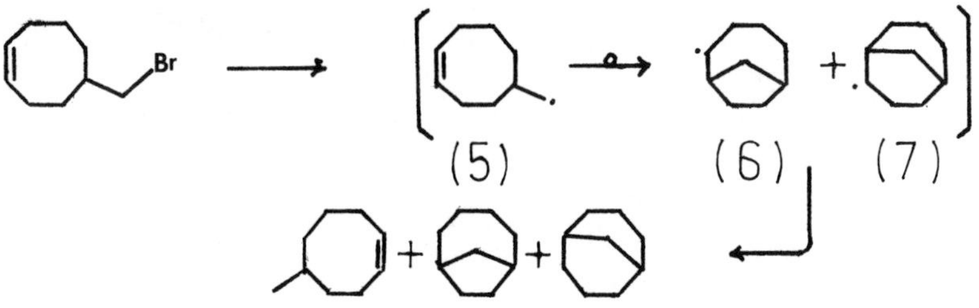

H. Fischer, H. Heimgartner (Eds.)
Organic Free Radicals
© Springer-Verlag Berlin Heidelberg 1988

The yields of the rearranged products are strongly dependent on temperature and $^nBu_3SnH$ concentration. Up to 70 - 80% rearranged products can be obtained. The rate constants ($\times 10^{-5}/s^{-1}$ at 25 °C) for cyclisation to (2), (6) and (7) have been determined as 1.0, 1.5 and 0.3 respectively in the usual way.[3,4]

The high yield of (4) was surprising as, of the three known conformers of the cycloheptene ring (11) - (13), only the highest energy boat, (13), offers the possibility of cyclisation. Using molecular mechanics and low temperature nmr, we have identified a fourth conformation (14) which also permits the rearrangement.

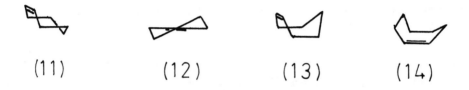

(11)   (12)   (13)   (14)

References

1. Beckwith ALJ and Ingold KU (1980) In: de Mayo P (ed) Rearrangements in Ground and Excited States. Academic Press, New York, ch.4, p.161
2 MacCorquodale F and Walton JC (1987) J Chem Soc Chem Commun 1456
3 Beckwith ALJ and Moad G (1980) J Chem Soc Perkin Trans 2 1083
4 Beckwith ALJ and Moad G (1974) J Chem Soc Chem Commun 472

CIDNP DURING THE OXIDATION
OF TERTIARY ALIPHATIC AMINES

S.A. Markarian

Department of Chemistry, Yerevan State
University, 375049, Yerevan,Armenia,USSR

Abstract: On the basis of the data obtained from our experiments and
other sources on the CIDNP effects observed during the photo and
thermal reactions of tertiary alkyl amines with electron-acceptor
compounds such as haloalkanes, peroxides, ketones etc. we present general radical mechanism for the two step dehydrogenization of amine.

U.V. studies revealed that tertiary amines (Et_3N, Bu_3N) and halocar-
bones (CCl_4, $CHCl_3$, $CBrCl_3$, C_2Cl_6) form weak complexes by electron
donor-acceptor interaction even in polar solvents (methanol, acetonitrile). Irradiation in the region of the charge transfer absorption
($270 \leqslant \lambda \leqslant 330nm$) in photo-CIDNP system leads to enhanced absorption
(A) for chloroform (in the case of C_2Cl_6 for C_2Cl_5H) and for terminal
olefinic protons of N, N diethylvinylamine (δ, 4.1p.p.m). From the
known free radical parameters and Kaptein's rules it follows that po-
larization is created in the intermediate radical pair $\overline{Et_2N\overset{\bullet}{C}HCH_3 \; \overset{\bullet}{C}Cl_3}$
$\overline{(or \; C_2Cl_5)}^S$. When aqueous NaOH was added to the methanol solutions be-
fore irradiation no CIDNP of the starting amine was observed. These
findings support that the previously formed ion-radical pair $Et_3N^{+\bullet}$
$CCl_4^{\overline{\cdot}}$ (or $C_2Cl_6^{\overline{\cdot}}$) transforms to the neutral radical pair by HCl elimination within nanoseconds. CIDNP kinetics for chloroform for the
reaction of triethylamine with CCl_4 was investigated. The rate cons-
tonts (K) and enhancement coefficients (E) were: $k = 3,2 \cdot 10^{-2} s^{-1}$
$E = 10$ (in methanol) and $k = 1,5 \cdot 10^{-2} s^{-1}$ $E = 9$ (in acetonitrile).
Triethylamine acts as a dehalogenated agent and in appropriate conditions can transform halocarbons into corresponding hydrocarbons; for
example by the following scheme: $CCl_4 {\rightarrow} CHCl_3 {\rightarrow} CH_2Cl_2 {\rightarrow} CH_3Cl {\rightarrow} CH_4$.

The reactions of tertiary amines with benzoyl peroxides occur through
formation of an unstable quarternary derivative (SN_2-mechanism)
$[\; R_3\overset{+}{N}O\overset{\bullet\bullet}{\underset{O}{C}}Ph \, PhCO_2^{-} \;]$ which reacts further to give both radical and
non-radical derived products. Enhanced CIDNP was observed for the

H. Fischer, H. Heimgartner (Eds.)
Organic Free Radicals
© Springer-Verlag Berlin Heidelberg 1988

acetic proton of benzoic acid during the reaction of triethylamine with benzoyl peroxide at $+120°C$ in excess of amine. These results together with the data obtained by Lawler, Roth, McLauchlan for the other oxidation reactions of tertiary aliphatic amines are well explained in frame of the following general mechanism

$$R_3N + AX \longrightarrow R_3N \cdots AX \, (R_3\overset{+}{N} \, AX^-) \longrightarrow \overline{R_3N^{\ddagger} \, AX^{\overline{\cdot}}} \,{}^{\ominus} \longrightarrow$$
$$\xrightarrow[-HA]{} R_2N\overset{\cdot}{C}HR' \, \overset{\cdot}{X}{}^{\ominus} \longrightarrow R_2NCH = CHR'' + HX$$

here AX is an electron-acceptor compound. We have also isolated the secondary product of oxidation of amine with various acceptors and identified it as a paramagnetic polymer with π-electronic conjugation

$$-(CH = \underset{NR_2}{\overset{|}{C}})_n - (CH = CH)_m^-$$

It was shown that this polymer forms from vinylamine and it is responsible for yellow-brown colouration of amine solutions.

CHARGE TRANSFER INTERACTION BETWEEN
MENADIONE AND NUCLEIC ACIDS. A CIDNP STUDY.

J.Marko , G.Vermeersch , V.Toebat , N.Febvay and A.Lablache-Combier#

Lab. de Physique , Faculté de Pharmacie , F-59045 LILLE-Cédex , FRANCE.
#Lab.de Chimie Organique Physique , associé à l'ENSCL , LA-CNRS 351 , Université de LILLE I , F-59655
VILLENEUVE D'ASCQ-Cédex , FRANCE.

2-Methyl-1,4-naphthoquinone or menadione (MQ) ,which is a component of vitamin K_3 , is known to photooxidize various pyrimidine bases from ADN upon exposure to UV light (1).It seems very likely that a pyrimidine cation radical should be formed as a primary intermediate in these photoreactions,as suggested by flash photolysis experiments (2).This cation radical should lead to the various photoproducts obtained through the sensitized photooxidation of thymidine (3) or 2'-desoxycytidine (4) by MQ. It should result from a charge transfer reaction between MQ in its triplet excited state and the pyrimidine.

Hence it seems that Photochemically Induced Dynamic Nuclear Polarization (Photo-CIDNP) would be a suitable tool for the study of the primary photoprocesses occuring with MQ and such substrates.
We have still used this technique to study similar charge transfer interactions between some photosensitizing drugs such as the phenothiazines (5) and synthetic water-soluble porphyrins (6) with several nucleobases.
We will report here the results for MQ and its bisulfite derivative in aqueous (D_2O) and alcoholic solutions (CD_3OD).

In the absence of substrate,polarized photoproducts are observed with the bisulfite in both solvents and with MQ in methanol.They will be compared with authentic samples isolated from the irradiated mixture.
For MQ in D_2O (with added DMSO-d_6 for solubilization) line broadening phenomena were detected on the starting material.It probably accounts for an energy exchange reaction between a triplet excited molecule (^3MQ) and a ground state one.

The formation of the photoproducts is quenched when the solutions are irradiated in the presence of cytosine or adenine.Simultaneously strong polarizations are observed on both nucleobases by ^1H and ^{13}C CIDNP.They should be consistent with the formation of the nucleobase radical cation (5).Further reactions of these species do not lead to any polarized photoproducts,even through extended irradiation (more than 3 min.) and only the back electron transfer step could be detected by CIDNP.
The quenching seems to be less efficient with the corresponding nucleosides and mononucleotides since the CIDNP effects on the latter,corresponding to the above mentioned transfer step,are only observed at the beginning of irradiation.Then the polarizations are similar to those obtained when MQ is irradiated alone in solution.

The situation is quite different when thymine is used as the nucleobase.Besides the effects which are described when MQ is irradiated in the absence of substrate polarized

H. Fischer, H. Heimgartner (Eds.)
Organic Free Radicals
© Springer-Verlag Berlin Heidelberg 1988

photoproducts,probably resulting from the thymine radical cation (4) are also observed less than 20 s after the light is turned on.Polarizations on thymine (enhanced absorption for the methyl protons) which are seen in the first 20s rapidly collapse.Hence the thymine radical cation should be more reactive than the cytosine or adenine ones.Identical results are obtained with thymidine monophosphate.

On the other hand no polarizations are observed whith guanine,whose presence do not modify the MQ behaviour.

References

1 Wagner JR , Cadet J , Fisher GJ (1984) Photochem. Photobiol. 40:589
2 Fisher GJ , Land EJ (1983) Photochem. Photobiol. 37:27
3 Decarroz C , Wagner JR , Van Lier JE , Murali-Krishna C , Riesz P , Cadet J (1986) Int. J . Radiat. Biol. 50:491
4Murali-Krishna C , Decarroz C , Wagner JR , Cadet J , Riesz P (1987) Photochem. Photobiol. 46:175
5 Marko J , Vermeersch G , Febvay N , Lablache Combier A (1985) Photochem. Photobiol. 42:213
6 Le Nouen D , Marko J , Vermeersch G , Febvay N , Lablache-Combier A , Perrée-Fauvet M , Gaudemer A (1988) Photochem. Photobiol. , to be published

GRIGNARD REACTION OF KETONES. ELECTRON AND SUBSEQUENT

R· TRANSFER PROCESSES TO KETONES

K. Maruyama and T. Katagiri

Department of Chemistry, Faculty of Science,
Kyoto University, Kyoto 606, Japan

Abstract: Grignard reactions of benzil, benzophenones, fluorenbnes, and xanthone
were investigated. Spectroscopic techniques such as ESR and stopped flow showed
presence of stable radicals in reacting solutions under suitable conditions.
The structures of radicals were proposed. Dependence of products distribution
(addition/reduction) on the concentrations and properties of solvents were confirmed. Such phenomena will be explained by the mechanism proposed by authors previously.

1. Structures of stable radicals in Grignard reactions

By selecting suitable conditions (concentration ratio of both Grignard reagent and
ketone, extremely dry and deoxygenated solvents) stable radicals were able to
observe in the Grignard reactions of ketones in almost all examined cases.

Fig. 1   Two stable radicals observed in the Grignard reaction of benzil
         and phenylmagnesium bromide

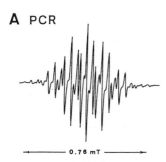
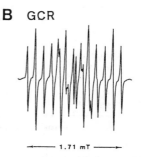

A PCR — 0.76 mT

B GCR — 1.71 mT

Fig. 2   Structure of a stable radical (corresponds A, PCR in Fig.1)

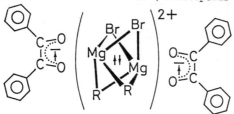

H. Fischer, H. Heimgartner (Eds.)
Organic Free Radicals
© Springer-Verlag Berlin Heidelberg 1988

2. Dependence of addition/reduction ratio upon concentrations both of ketone and Grignard reagent, and properties of solvents.

Table 1

Ratio of addition, and reduction products in the Grignard reaction depending upon the various mixing ratios of the reagents in THF at room temperature.

ketone(M)	Grignard reagent(M)	G/K[a]	conversion(%)[b]	addition/reduction
benzophenone	EtMgBr[d]			
(0.079)	(0.157)	2.00	99	0.214
(0.126)	(0.157)	1.25	98	0.144
(0.157)	(0.157)	1.00	96	0.138
(0.188)	(0.157)	0.83	78	0.144
(0.314)	(0.157)	0.50	50	0.139
benzophenone	n-BuMgBr[c]			
(0.077)	(0.153)	2.00	99	0.176
(0.122)	(0.153)	1.25	99	0.111
(0.153)	(0.153)	1.00	98	0.142
(0.184)	(0.153)	0.83	84	0.114
(0.306)	(0.153)	0.50	49	0.116

a) The ratio of initial concentrations of Grignard reagent and ketone; G/K = ["RMgBr"]*/[ketone]*. b) Conversion yield based on consumed ketone. c) Ratios of addition/reduction were determined both by weight of two isolated products and by ^{1}H-NMR of product mixture. d) Ratios were determined by ^{1}H-NMR.

Table 2.

Solvent Effect on the ratio of addition and reduction products in the Grignard reaction at room temperature. (RMgBr/Ketone = 2.0)

ketone(M)	RMgBr(M)	solvent[a]	addition/reduction[b]
fluorenone (0.079)	EtMgBr (0.159)	DEE	3.81 ± 0.20
fluorenone (0.079)	EtMgBr (0.157)	THF	2.58 ± 0.18
benzophenone (0.079)	EtMgBr (0.159)	DEE	1.55 ± 0.09
benzophenone (0.079)	EtMgBr (0.157)	THF	0.214 ± 0.018
benzophenone (0.077)	n-BuMgBr (0.153)	DEE	0.366 ± 0.039
benzophenone (0.077)	n-BuMgBr (0.153)	THF	0.176 ± 0.010

a) DEE = diethyl ether. b) Ratios of addition/reduction were determined by ^{1}H-NMR.

Table 3   Products in the Grignard reactions of benzil.

Grignard reagent (M)		solvent	products (%)		
			CAP	OAP	benzoin
EtMgBr	(0.05)	THF	62	38	0
	(0.10)	THF	56	44	0
	(0.15)	THF	57	43	Tr.
	(0.50)	THF	54	46	Tr.
EtMgBr	(0.10)	DEE	40	60	Tr.
PhMgBr	(0.05)	THF	100	Tr.	0
	(0.10)	THF	97	3	0
	(0.15)	THF	95	5	0
	(0.50)	THF	90	9	0
PhMgBr	(0.10)	DEE	99	Tr.	0
MeMgBr	(0.05)	THF	100	0	0
	(0.10)	THF	100	0	0
	(0.15)	THF	100	0	0
	(0.50)	THF	99	0	0

Benzil / Grignard reagent = 1.0, mean value of three runs, at room temperature, 3 days.

As shown in Table 1 addition/reduction ratio changes with relative concentrations of reactants. Similarly, addition/reduction ratio changes remarkably with solvents. THF and EtOEt were used as solvents. (see Table 2)

3. Change of C-addition product (CAP)/O-addition product in the reaction of benzil depending upon species of Grignard reagents, its concentrations, and properties of solvents.

As shown in Table 3 in the reaction of benzil C-addition product/ O-addition product(OAP); (CAP)/(OAP) changed with change of species of Grignard reagents, concentration of reactants, and solvents. These results will be discussed by the mechanism proposed by us.

References

K.Maruyama and T.Katagiri, J. Amer. Chem. Soc., 108, 6263(1986); K.Maruyama and T.Katagiri, Chem. Lett., 1987, 731, 735 ; K.Maruyama and T.Katagiri, Chem. Lett., 1986, 601 ; K.Maruyama and T.Katagiri, J. Phys. Org. Chem., 1, 21 (1988)

RADICAL IONS AND PHOTOCHEMICAL CHARGE TRANSFER PHENOMENA IN ORGANIC CHEMISTRY

J.Mattay

Institut für Organische Chemie, RWTH Aachen,
Prof.-Pirlet-Str.1, D-5100 Aachen

Photochemical excitation of electron-acceptor(A) or electron-donor(D) substrates leads to well-defined changes in their redox properties, i.e. A(D) becomes even a stronger acceptor (donor) after excitation [1]. In general, the feasibility of producing radical ions by photoinduced electron transfer can be predicted by means of the Weller equation [1,2]. Furthermore the use of polar solvents, the exploitation of the special salt effect (eq.1) and fast chemical reactions (Scheme) may overcome the back electron transfer, which would only lead to energy wastage [1].

$$A + D \xrightarrow{\ h\nu\ } (A^{\overline{\cdot}}D^{\underline{+}}) \xrightarrow{\ Li^{+}\ } (A^{\overline{\cdot}}Li^{+}) + D^{\underline{+}} \qquad (1)$$

In addition we have used the Weller equation to estimate the degree and the direction of charge transfer and we have correlated the free enthalpies of electron transfer (ΔG) with the selectivities of photoreactions between A and D [1,3].

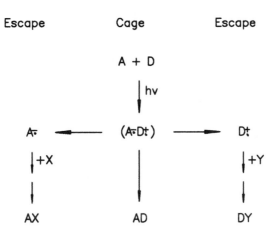

H. Fischer, H. Heimgartner (Eds.)
Organic Free Radicals
© Springer-Verlag Berlin Heidelberg 1988

Here we will focus on photoreactions involving charge separation according to the simplified presentation given in the scheme. One part will deal with donor–acceptor reactions, which are often cage processes. In the second part we will discuss examples resulting from escape, such as cycloadditons of olefin radical cations. Exeptions from this categorization will be discussed as well, especially the role of contact ion pairs (CIP) and solvent separated ion pairs (SSIP) in radical cation catalyzed Diels Alder reactions [4]. Finally, recent results on new types of electron transfer senzitizers and homogeneous metal catalysis of intramolecular cycloadditions involving charge separation will be discussed.

[1]   Mattay, J. *Angew.Chem.* 1987,*99*,849; *Angew.Chem.Int.Ed.Engl.* 1987,*26*,825.
[2]   Weller, A. *Z.Phys.Chem.N.F.* 1982,*133*,93.
[3]   Mattay, J. *J.Photochem.* 1987,*37*,167.
[4]   Mattay, J. *Nachr.Chem.Tech.Lab.* 1988,*36*,376.

CYCLISATION OF γ-ARYLALKANOLS VIA ARYL RADICAL-CATION AND ALKOXYL RADICAL INTERMEDIATES

B.C. Gilbert[1] and C.W. McCleland*[2]

1. Dept. of Chemistry, University of York, Heslington, York YO1 5DD, U.K.

2. Dept. of Chemistry, University of Port Elizabeth, P.O. Box 1600, Port Elizabeth 6000, South Africa

A comparison is made of the cyclisation reactions of the aryl radical-cations 2 and alkoxyl radicals 4 derived from the γ-arylalkanols 1. This system is of interest since interconversion of the radical-cation and alkoxyl radical intermediates is possible (Scheme). Our objective has been to gain insight into the factors influencing their cyclisation and interconversion.

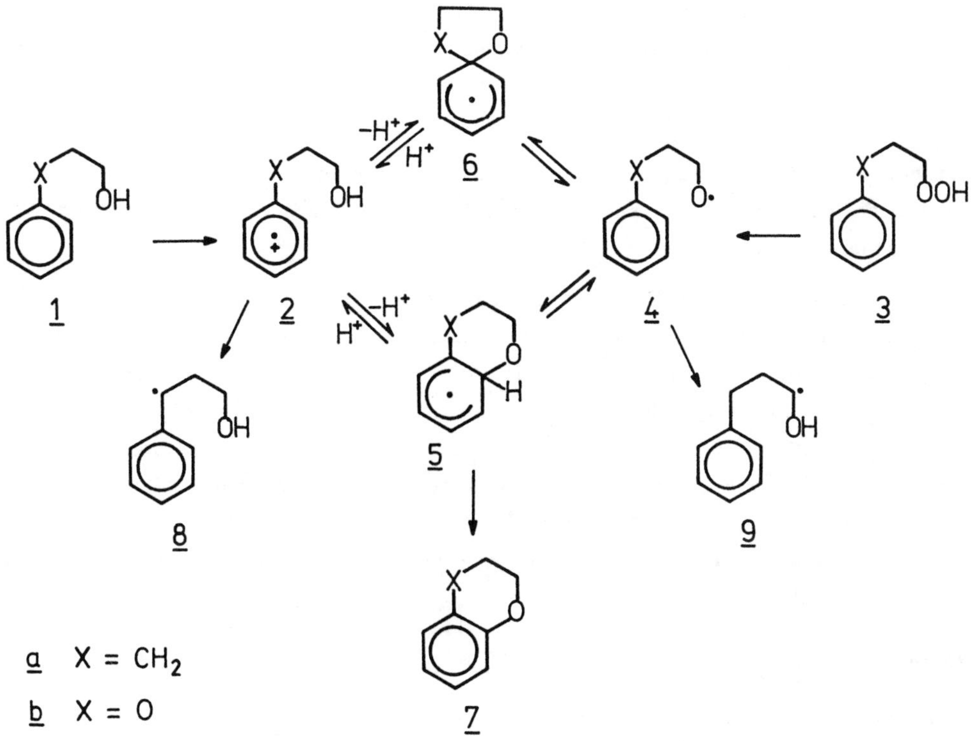

a   X = CH₂

b   X = O

SCHEME

H. Fischer, H. Heimgartner (Eds.)
Organic Free Radicals
© Springer-Verlag Berlin Heidelberg 1988

The radical-cations 2 were obtained by reacting the alcohols 1 with $SO_4^{\cdot-}$ generated from $S_2O_8^{2-}$ either thermally or by metal-ion (Ti^{3+} or Fe^{2+}) reduction (1), with or without added Cu^{2+}. Yields of cyclic product 7 were very low ($<6\%$) except for $7a$ when formed in the presence of Cu^{2+} (62%).

The alkoxyl radicals 4 were generated by metal-ion (Ti^{3+} or Fe^{2+}) reduction of the hydroperoxides 3 (2), also with or without added Cu^{2+}. Little cyclisation ($< 4\%$) was evident in the absence of Cu^{2+}, but significant yields of both $7a$ (70%) and $7b$ (27%) were obtained in its presence.

These results are consistent with reversible cyclisation of radical-cation $2a$ which affords $7a$ only when Cu^{2+} is present to oxidise the intermediate radical $5a$. However, in the absence of Cu^{2+}, some conversion of $5a$ or $6a$ to the alkoxyl radical $4a$ is proposed to account for the formation of various side-products, including ethyl benzene. In contrast, the radical-cation $2b$ does not cyclise, probably as a result of mesomeric stabilisation imparted by the oxygen atom.

The obtainment of 7 from the alkoxyl radicals 4 only in the presence of Cu^{2+} is also due to reversible cyclisation (3). The yields of 7 were reduced at lower pH, possibly as a result of competing conversion of intermediates 5 to radical-cations 2. The generally lower yield of $7b$ relative to $7a$ is also due to the more facile formation of $2b$ and its failure to cyclise.

Although no products unique to the 1,5-cyclisation route could be identified, it cannot be excluded since such intermediates are known to rearrange to their 1,6-analogues (3).

An e.s.r. study of the radical-cation $2a$ generated from the alcohol $1a$ with $SO_4^{\cdot-}$ or $H\dot{O}/H^+$ (4) was conducted. The former method was not illuminative owing to rapid oxidation (5) of any α-hydroxy alkyl radical 9 by $S_2O_8^{2-}$; only the benzylic radical 8 was observed. However, when reacted with $H\dot{O}/H^+$, alcohol $1a$ displayed radical 9 at pH 2 and radical 8 at lower pH. These observations accord with the formation of radical-cation $2a$ which cyclises and then ring-opens to the alkoxyl radical $4a$. The radical 9 is not formed from $1a$ by H-abstraction since the analogous radical was not obtained when the methyl ether of alcohol $1a$ was similarly treated.

REFERENCES

1 Minisci F, Citterio A, Giordano C, Acc Chem Res 1983, 16, 27.
2 Kochi, J K in Kochi J K (ed) (1973) Free Radicals Vol 1, Wiley, New York.
3 Goosen A, McCleland C W, S Afr J Chem 1978, 31, 67.
4 Davies M J, Gilbert B C, McCleland C W, Thomas C B, Young J, J Chem Soc, Chem Commun 1984, 966.
5 Davies M J, Gilbert B C, Norman R O C, J Chem Soc, Perkin Trans 2 1984, 503.

THE QUESTION OF ADDITIVITY OF SUBSTITUENT EFFECTS ON RADICAL STABILISATION

R. Merényi, Z. Janousek and H.G. Viehe

Laboratoire de Chimie Organique, Université de Louvain
1 Pl. L. Pasteur B-1348 Louvain-la-Neuve.

By measuring the hyperfine splittings (h.f.s.) of α-protons in benzylic radicals carrying para substituents Arnold has established the σ_α^o-scale of radical sta-bilisation (1). This scale correlates well with the Relative Radical Stabilisation (RRS) scale, which gives the average values of Radical Stabilization Energies determined by various methods (2). Furthermore it is known that the cumulative interaction of substituents on spin delocalization follows Fischer's relation (3) which is not based on arithmetic additivity. By analogy Radical SE-s should also follow this rule and result in values smaller than additive. For the expression of deviations from Fischer's equation Rhodes and Roduner proposed a substituent interaction parameter Δ_{XY} for two substituents X and Y (4). Positive values mean synergy of the effect of the two substituents, negative ones antagonism.

$$A_{XY}/A_{HH} = (1-\Delta_X) \ (1-\Delta_Y) \ (1-\Delta_{XY})$$

In this formula A_{XY} and A_{HH} are proton- and muon-electron coupling constants respec-tively.We think that activation barriers of isomerisation or bond dissociation ener-gies can also be used to calculate Δ_{XY} values by the same formula in appropriate systems. A compilation of some (Δ_{XY}.100) values derived from results of several authors is given in Table 1. These values are calculated from ESR data on substitu-ted benzylic (5) and propargylic (6) radicals, from muon coupling constants of Mu-adducts to disubstituted benzene derivatives (4) and from relative rates of rear-rangement of methylene-cyclopropanes (substituted on the ring by COOMe and X in phenylogous position) (7) and furthermore from our measurements of meso to dl isome-risation kinetics ($\Delta G^{\neq}$) of the benzylic radical dimers **1a** to **1b**.

1a (meso)   **1b** (dl)

H. Fischer, H. Heimgartner (Eds.)
Organic Free Radicals
© Springer-Verlag Berlin Heidelberg 1988

When the substituents X in these dimers are varied the steric effect on the C-C bond to be homolysed remains constant as well as the σ-interactions between polar substi-tuents in geminal or vicinal position in order to minimize proximity effects.

Table 1 Substituent interaction parameter ($\Delta_{XY} \cdot 100$) of radical stabilisation

X	Y	Ref.(5)	Ref.(6)	Ref.(4)	Ref.(7)	1a⇌1b
CN	CN	-12.5				-2.4
CN	COOMe	-12.0				-2.1
COOR	COOR	-12.0			-0.8	
SOMe	COOMe				-0.6	
SO2Me	COOEt				-0.6	
OMe	OMe	-13.6		-3.4		
Me	NH2		-7.7			
Me	OH		-3.3			
OMe	CN	+16.0		+5.17		+1.0
CN	OMe			+6.98		
SR	CN	+13.0				-0.3
NH2	CN	+ 9.6				+4.2
N(SiMe3)2	COOEt		+12.8			
OH	COOEt		+ 7.4			
OMe	COOET				+2.6	
SMe	COOEt				+0.8	

The substituent interaction parameters in Table 1 show that in the case of two substituents of like polarity antagonism is always observed, whereas the *captodative* effect for substituents of opposite polarity is supported by Δ_{XY} values which are practically all positive, indicating synergy.

References:
1 Dust JM, Arnold DR (1983) J Am Chem Soc 105:1221,6531
2 Merényi R, Janousek Z, Viehe HG (1986) p 301 in Merényi R, Janousek Z, Viehe HG (eds) Substituent Effects in Radical Chem. NATO ASI Series 189. Reidel, Dordrecht
3 Fischer H (1964) Z Naturforsch A 19:866
4 Rhodes CJ, Roduner E (1988) Tetrahedron Letters 29:1437
5 Korth HG, Lommes P, Sustmann R, Sylvander L, Stella L (1987) Nouv J Chim 11:365
6 MacInnes I, Walton JC (1987) J Chem Soc Perkin Trans 2 1077
7 Creary X, Merhsheikh-Mohammadi ME (1986) J Org Chem 51:2664

SPIN DELOCALISATION IN HETEROCYCLIC CAPTODATIVE RADICALS

R. Merényi, C. Nootens, H.G. Viehe

Laboratoire de Chimie Organique, Université de Louvain

1 Pl. L. Pasteur B-1348 Louvain-la-Neuve.

For comparison with non cyclic carbon centered radicals with captodative (*cd*) substitution (1,2), such as **1**, we studied by ESR heterocyclic radicals **2–5**; these have the polar substituents incorporated into the cycle and may furthermore contain a double bond (**4,5**).

1 to 4	a	b	c	d	e	f
X	S	S	O	O	S	S
R	COOMe	CH₃	CH₃	CH₃	H	H
R'	H	H	H	CH₃	H	SEt
R"	H	H	H	H	CH₃	CH₃

Table 1. ESR data of radicals **1** to **4** (a^H values in Gauss).

Radical	X	g	$a_\beta^{CH_3}$	a_γ^{SCH}	a_γ^{OCH}	a_γ^{COCH}	a_α^{H}	$a_\delta^{COOCH_3}$
1a	S	2.0061		6.38		0.75		0.66
2a	S	2.0079		5.90		1.92		0.65
3a	S	2.0062		5.50		0.50		0.40
2b	S	2.0078	13.0	4.60		3.59		
2c	O	2.0060	13.75		2.60	7.35		
2d	O	2.0054	13.63		2.88	7.48		
						6.78		
4b	S	2.0060	13.2	1.83		1.05		
4d	O	2.0046	13.78			4.76		
2e	S	2.0069		4.13			13.0	
2f	S	2.0064		1.77				
						1.73		

H. Fischer, H. Heimgartner (Eds.)
Organic Free Radicals
© Springer-Verlag Berlin Heidelberg 1988

In this work we examine whether a cyclic structure because of restricted confor-mational freedom shows increased spin delocalisation, and also the influence of cyclic conjugation.

The radicals have been generated either by hydrogene abstraction using di-t-butylperoxide cleavage by UV or by thermal or photochemical dissociation of the radical dimers. The ESR data are summarized in table 1.The spectra of **5** will be discussed elswhere (3).

As a basis for the evaluation of the delocalisation the following data can be used: 1) the β-methyl couplings, which are known to be proportionnal to the spin density on the radical center (4), 2) the γ-couplings like SCH_3 and the cyclic SCH_2, OCH_2, $COCH_2$ and the δ-$COOCH_3$ couplings. Comparing the hyperfine splittings related to the same function in the systems **1** to **4** gives information on the extent of delocali-sation.

The spectra of the cyclic systems **2a** and **3a** appear comparable to that of the open chain structure **1a**, with probably conformationally diminished delocalisation in the six-membered ring.

The difference in the β-methyl couplings of the O- and S-heterocycles (**2b-d**) ref-lects the higher delocalisation by the thioalkyl substitution, but less than expec-ted from data on monosubstitution.In the O-heterocyclic radicals **2c** and **d** spin den-sity seems to be more displaced to the COR substituent than in the thio-analogue **2b**.

The insaturation in **4b** and **d** do not lead to enhanced delocalisation.

From ESR data on monosubstituted radicals the amount of synergism in these disubsti-tuted radicals can be estimated, it confirms the *cd* -postulate.

References:

1 Viehe HG, Merényi R, Stella L, Janousek Z (1979) Angew Chem 91:982; Angew Chem Internat Ed Engl 18:917

2 Viehe HG, Janousek Z, Merényi R, Stella L (1985) Acc Chem Res 18:148

3 Ghosez A, (1987) PHD Thesis, Louvain-la-Neuve

4 Fischer H, (1964) Z Naturforsch A 19:866

ß-FRAGMENTATION OF ALKYL RADICALS.
POLAR AND STERIC EFFECTS.

K. Klenke and J.O. Metzger

Department of Chemistry, University of Oldenburg
P.O.Box 25 03, D-2900 Oldenburg

Radical fragmentation is the reverse of radical addition. The importance of polar and steric effects to addition reactions is well known[1]. We report here the influence of polar and steric effects on the selectivity of ß-scission of alkyl radicals.

Polar effects: Alkyl radicals **2** have been generated by addition of cyclohexyl to alkenes **1**[2]. Methyl vs. ethyl have been eliminated from **2** to give products **3** and **4** (E- and Z-isomers). Some results are summarized in Table 1.

$$Et\text{-}\underset{\underset{X}{|}}{\overset{\overset{Me}{|}}{CH}}\text{-}C\text{=}CH_2 \;+\; \underline{c}\text{-}C_6H_{11}\overset{\bullet}{} \;\rightleftharpoons\; Et\text{-}\underset{\underset{X}{|}}{\overset{\overset{Me}{|}}{CH}}\text{-}\overset{\bullet}{C}\text{-}CH_2\text{-}(\underline{c}\text{-}C_6H_{11})$$

1   **2**

- Me$\overset{\bullet}{}$ - Et$\overset{\bullet}{}$

$$Et\sim CH=C\overset{\nearrow CH_2\text{-}(\underline{c}\text{-}C_6H_{11})}{\searrow X} \qquad Me \sim CH=C\overset{\nearrow CH_2\text{-}(\underline{c}\text{-}C_6H_{11})}{\searrow X}$$

3   **4**

a :	**b** :	**c** :	**d** :
X : COOMe	Ph	p-MeO-Ph	p-Cl-Ph

Tab. 1: ß-Scission of alkyl radicals **2** at 350°C.

Radical **2**	Yield[a] [mol%]	**4/3**	s[b]	$\Delta E_{a;4-3}$[c] [kJ/mol]	A_4/A_3[c]
a	14.7	17.8	1.25	-18.2 ± 2.0	0.6 ± 0.2
b	6.0	7.3	0.86	- 4.5 ± 0.5	3.0 ± 0.3
c	7.4	8.9	0.95	- 8.4 ± 0.6	1.6 ± 0.2
d	9.0	9.4	0.97	- 7.2 ± 0.8	2.4 ± 0.5

a) Σ products **3** and **4**. b) selectivity log [**4**] / [**3**]. c) 300 - 450°C.

H. Fischer, H. Heimgartner (Eds.)
Organic Free Radicals
© Springer-Verlag Berlin Heidelberg 1988

The competitive rate data of Table 1 clearly indicate the influence of polar effects on fragmentation reactions. Relative rates of cleavage of the more nucleophilic ethyl are increased by electron withdrawing substituents X in radical **2**. The more reactive radical **2a** is fragmented with higher selectivity than the less reactive radical **2b**.

Steric effects: Radical **6** decomposes by two routes to give methyl and alkenes **7** and **8**. Relative rates for the decomposition have been obtained from analysis of the olefinic products. (Table 2). On a per methyl basis methyl has been eliminated faster $(k_2/k_1 > 1)$ from the less substituted carbon atom in radical **6** to give the less stable alkene.

We explain this surprising result as follows: ß-scission is influenced by steric effects in the same way as the reverse radical addition. Radicals add faster to the less substituted end of an alkene due to steric effects. The same must be true for ß-scission.

$$Me_{(3-m)}H_mC-C-CMe_{(3-n)}H_n \quad + \quad \underline{c}\text{-}C_6\dot{H}_{11} \quad \longrightarrow \quad Me_{(3-m)}H_mC-\dot{C}-CMe_{(3-n)}H_n$$

with CH_2 (in **5**) and $CH_2\text{-}(\underline{c}\text{-}C_6H_{11})$ (in **6**)

5 → $-\dot{Me}$ k_1 / k_2 $-\dot{Me}$ **6**

$$Me_{(2-m)}H_mC=C-CMe_{(3-n)}H_n$$
$$CH_2\text{-}(\underline{c}\text{-}C_6H_{11})$$
7

$$Me_{(3-m)}H_mC-C=CMe_{(2-n)}H_n$$
$$CH_2\text{-}(\underline{c}\text{-}C_6H_{11})$$
8

a: m=0;n=2;   **b**: m=0;n=1,   **c**: m=1;n=2.

Tab. 2: Selectivity of the ß-scission of alkyl radicals **6** at 400°C.

Radical **6**	[**7**] mol%	[**8**] mol%	k_2/k_1 [a]	S [b]
a	57	43	2.3	0.36
b	52	48[c]	1.4	0.15
c	62[c]	38	1.2	0.08

[a] Relative rate on a per methyl basis; [b] $S = \log k_2/k_1$; [c] E- and Z-products.

1 J.M. Tedder (1982) Angew. Chem. **94**, 433; Angew. Chem. Int. Ed. Engl. **21**, 401; B. Giese (1983) Angew. Chem. **95**, 771; Angew. Chem. Int. Ed. Engl. **22**, 753
2 J. Hartmanns, K. Klenke und J.O. Metzger (1986) Chem. Ber. **119**, 488

THE DECAY KINETICS OF TRIPLET-TRIPLET FLUORESCENCE SYSTEMS. THE M-XYLYLENE BIRADICALS.

A. Després, V. Lejeune and E. Migirdicyan

Laboratoire de Photophysique Moléculaire du CNRS
Bâtiment 213, Université Paris-Sud, Orsay, FRANCE

The UV photolysis of methyl substituted benzenes in Shpolskii matrices at 5-20K gives rise to benzyl-type monoradicals together with m-xylylene biradicals. These fragments result from the photodissociation of CH bonds on one or two different methyl groups of the parent molecule. The mono-and biradicals can be distinguished by the energies and vibrational structures of their quasi-line fluorescence and excitation spectra obtained by site-selective laser experiments.(1)

Theory predicts that m-xylylene biradicals (C_{2v}) have a triplet ground state of B_2 symmetry (this has been confirmed by ESR experiments (2) for m-xylylene and for 3-methyl-m-xylylene) and two close-lying triplet excited states having A_1 and B_2 symmetries. Calculations predict the presence of three and four singlet states between the ground T_0 and the first excited triplet T_1.(1)

The fluorescence decays of benzyl-type monoradicals and of m-xylylene biradicals trapped in Shpolskii matrices have been measured in the same experimental conditions. While the decay is a single exponential for monoradicals, it is a sum of two exponential functions for biradicals in the whole temperature range between 10 and 77K. The lifetimes τ_1 and τ_2 of the two components determined for the biradicals produced from perprotonated and perdeuterated mesitylene (M-h_{10} and M-d_{10}) and durene (D-h_{12} and D-d_{12}) are listed in Table 1.

Table 1 - Observed lifetimes of the two components determined from the fluorescence decays of biradicals produced from mesitylene and durene.

	M-h_{10}	M-d_{10}	D-h_{12}	D-d_{12}
τ_2 (ns)	34 $\pm$ 5	50 $\pm$ 8	115 $\pm$ 15	220 $\pm$ 20
τ_2 (ns)	200 $\pm$ 15	260 $\pm$ 20	680 $\pm$ 70	1220 $\pm$ 150

H. Fischer, H. Heimgartner (Eds.)
Organic Free Radicals
© Springer-Verlag Berlin Heidelberg 1988

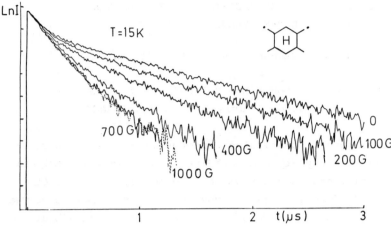

Fig 1 - The magnetic field effect on the fluorescence decays of $D-H_{12}$ in n-hexa-at 15 K

In the presence of a magnetic field, the fluorescence decay of monoradicals remains unchanged, while that of biradicals is significantly altered. The decay curves of biradicals $D-h_{12}$ in the field range 0-1000G are presented in figure 1. The analysis of these curves indicate that the lifetime of the long component decreases drastically while that of the short component remains practically unchanged.

One possible interpretation of these results is to attribute the biexponential decay to the fluorescence from two individual sublevels of the lowest excited triplet level which is the emitting state of the m-xylylene biradicals. This requires that the emission from triplet sublevels be faster than the spin-lattice relaxation between the different sublevels. This condition is probably satisfied for the 30 - 1200 ns fluorescence of m-xylylene biradicals.

Spin-orbit coupling is probably very small in m-xylylene biradicals. Consequently the radiative decays from the three T_1 sublevels are expected to be nearly equal. Spin-orbit coupling induces intersystem crossing. This radiationless transition in competition with the radiative decay is probably responsible for the large difference in the lifetimes of the short and long components.

In the presence of magnetic field, the zero-field wave functions of the T_1 sublevels will be mixed by the Zeeman term in the hamiltonian. As a result of this mixing, the long-lived zero-field level will be gradually depopulated by intersystem crossing, as the field strength increases.

References
(1) Lejeune V, Després A, Migirdicyan E, Baudet J and Berthier G (1986) J Am Chem Soc 108, 1853
(2) Wright B, Platz M S (1983) JACS 105, 628

ESR SPECTROSCOPY OF SHORT-LIVED RADICAL
PAIRS IN SOLUTIONS

Yu.N. Molin

Institute of Chemical Kinetics and Combustion,
630090, Novosibirsk, USSR

Short-lived radical pairs are, presently, the subject of permanent
scientific attention caused by both the radical pair chemical transformation proper, and their specific role in the appearance of such
phenomena as the chemically-induced polarization of electrons and
nuclei, the effect of external magnetic field and the resonance microwave (mw) fields on radical reaction, magnetic isotopic effect (see
e.g. (1)). Unfortunately, the ESR spectra of radical pairs in solutions cannot be detected by means of a conventional ESR technique
due to their low stationary concentration. The present review is concerned with more sensitive methods based on the radical pair spin
state modulation, and, hence, its reaction under the action of the
resonance mw radiation, and the intrinsic interactions of the magnetic nature in pair radicals. In the broad sense, they may be referred
to the methods based on the reaction yield detected magnetic resonance
(2).

Optically detected ESR (OD ESR): The recombination of singlet geminate radical-ion pairs resulting from radiolysis leads to the formation of singlet-excited luminescing molecules. These pair ESR spectra
are optically detected from the fluorescence intensity decrease at
the pumping of pair partner ESR transitions, which favours the sing-
let-to-triplet pair transformation. The apparatus allowing the OD ESR
spectra to be registered under the stationary irradiation from radio-
active source (3) and X-rays (4), and under the pulsed irradiation
of electron accelerator (5) have been developed. The spectra of radical cations of alkanes, olefins, and alkylamines, as well as radical-
anions of fluorobenzene derivatives have been detected by the OD ESR
method. The dimerization of olefinic and aromatic radical-cations,
and the charge transfer reactions involving radical-ions, have been
studied (see, e.g. (6-8)).

Spin-trapping technique: The ESR pumping affects the recombination of

H. Fischer, H. Heimgartner (Eds.)
Organic Free Radicals
© Springer-Verlag Berlin Heidelberg 1988

geminate radical pair, and, hence, its ESR spectrum may be determined by measuring the product yield of either its recombination or
the reaction of radicals escaping into the bulk. The technique has
been realized based on the scavenging of the escaped radicals by nitroxyl spin traps (9). The ESR spectra of radical pair generated photochemically in micelles have been obtained by this method.

Stimulated nuclear polarization: The ESR pumping of spin-correlated
radical pairs induces an additional nuclear polarization in the products of radical reactions, called a stimulated nuclear polarization
(10). The ESR spectra of radical pairs generated in solutions photochemically and containing semiquinone (10) and phenyl (11) radicals
have been detected by means of this technique.

Quantum beats: The spin-correlated radical pair undergoes the singlet-
triplet oscillations governed by hyperfine interactions, and the g-
factor difference of pair partners. These oscillations (quantum beats)
have been detected in the time-resolved luminescence experiments (12).
The hyperfine coupling constants (12), and besides the g-factors and
linewidths (13) of a series of radical-cations generated by ionizing
radiation have been obtained from the quantum beats.

References:

1 Salikhov KM, Molin YN, Sagdeev RZ, Buchachenko AL (1984) Spin polarization and magnetic effects in radical reactions. Elsevier,
   Amsterdam Oxford New York Tokyo
2 Frankevich EL, Kubarev SI (1981) In: Clarke (ed) Triplet state OD
   MR spectroscopy. John Willey and Sons
3 Anisimov OA, Grigoryants VM, Molchanov VK, Molin YN (1979) Chem
   Phys Lett 66:265
4 Molin YN, Anisimov OA (1983) Rad Phys Chem 21:77
5 Trifunac AD, Smith JP (1980) Chem Phys Lett 73:94
6 Melekhov VI, Anisimov OA, Veselov AV, Molin YN (1986) Chem Phys
   Lett 127:97
7 Werst DW, Trifunac AD (1988) J Phys Chem (in press)
8 Saik VO, Anisimov OA, Lozovoy VV, Molin YN (1985) Z Neturforsh
   40a:239
9 Okazaki M, Shiga T (1986) Nature 323:240
10 Bagryanskaya EG, Grishin YA, Avdievitch NI, Sagdeev RZ, Molin YN
   (1986) Chem Phys Lett 128:162
11 Bagryanskaya EG, Grishin YA, Sagdeev RZ, Molin YN (1985) Chem Phys
   Lett 114:138
12 Anisimov OA, Bizyaev VL, Lukzen NN, Grigoryantz VM, Molin YN (1983)
   Chem Phys Lett 101:131
13 Veselov AV, Melekhov VI, Anisimov OA, Molin YN (1987) Chem Phys
   Lett 136:263

STEREOCHEMICAL EFFECT ON
INTRAMOLECULAR SUBSTITUTIONS
IN INDUCED DECOMPOSITION OF
ALLYLIC PERCARBONATES

X. Lubeigt, E. Montaudon and B. Maillard

Université de Bordeaux I, 351, cours de la Libération, F-33405 TALENCE-Cédex

(France).

Abstract : Free radical additions of methylene chloride to unsaturated O,O-t-butyl O-allylic percarbonates showed the existence of a stereochemical effect on intramolecular substitution at the O-O bond of the percarbonate.

In a recent serie of articles (1), we described the synthetic potential of the induced homolytic decomposition of unsaturated peroxy-compounds. This reaction was found to be efficient as an access to cyclic esters and ethers :

$$Z^{\cdot} + \begin{cases} H_2C=CH\sim\sim\sim OOtBu \longrightarrow Z-CH_2-CH \underset{C}{|} \\ \\ H_2C=CH\sim\sim\sim CO_3tBu \longrightarrow Z-CH_2-CH \underset{O-C=O}{|} \end{cases} + \;\; tBuO^{\cdot}$$

We undertook the study of the influence of various factors on the course of this reaction :
- length of the chain linking the unsaturation and the peroxydic group
- effects of substituents on the yield of heterocycle (influence on the competition intramolecular homolytic substitution ($S_H i$) and hydrogen transfer from the solvent ZH)
- effect of substituents at the double bond on the size of the heterocycle (influence on the régiochemistry of the addition)
- nature of the unsaturation
- nature of the atom in α to the perester function.

To complete this work we started a study on the effects of substituents on the stereochemistry of the $S_H i$ reaction. We will describe here preliminary results obtained in the free radical addition of methylene chloride to O,O-t-butyl O-allylic percarbonates. According to previous mechanistic studies this reaction can be described by the following equation :

H. Fischer, H. Heimgartner (Eds.)
Organic Free Radicals
© Springer-Verlag Berlin Heidelberg 1988

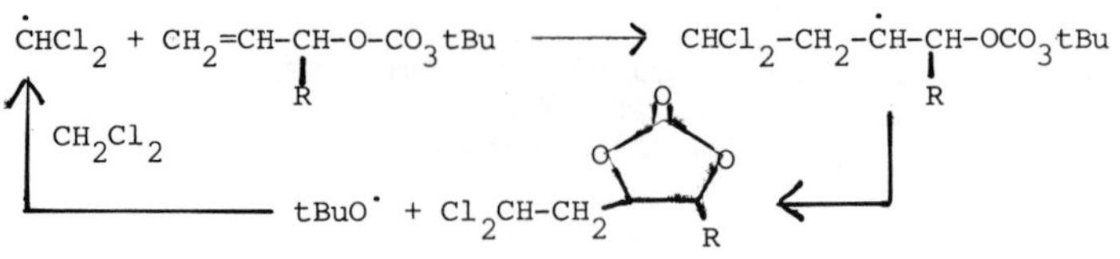

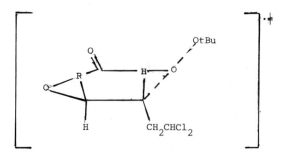

$$\overset{\cdot}{C}HCl_2 + CH_2=CH-CH-O-CO_3tBu \longrightarrow CHCl_2-CH_2-\overset{\cdot}{C}H-CH-OCO_3tBu$$

Table 1 summarizes the results obtained for such reactions with various R groups.

Table 1. Free radical addition of methylene chloride to unsaturated percarbonates (a)

R	Yield (%) (b)	cis/trans
CH_3	52	25/75
CH_3CH_2	64	23/77
$(CH_3)_2CH$	61	20/80
$(CH_3)_3C$	60	7/93

(a) CH_2Cl_2/Percarbonate/Benzoyl peroxide : 50/1/0,1 ; 80°C ; 24 h
(b) relative to peroxide

One can see that there is a stereochemical effect in all cases. Increasing the bulk
of the substituent R favours the predominance of the trans isomer but only significantly when all the hydrogen atoms of the carbon linked to the cycle are replaced by
methyl (R = t-Bu). This stereochemical effect can be understood by looking at the
interaction of the two substituents (R and CH_2CHCl_2) in the transition state :

Such a result motivates us to further investigate stereochemical effects in the formation of heterocycles by intramolecular homolytic substitution of carbon centered
radicals at the O-O bond.

Literature

(1) M.J. Bourgeois, E. Montaudon, B. Maillard, Bull. Soc. Chim. Belg., 97, 255,
    (1988) and precedent papers of this series.

GAS PHASE OXYGENATION OF BENZENE
DERIVATIVES WITH O(^{3}P) ATOMS
PRODUCED BY MD OF N$_2$O

Veronica M. Sol, Robert Louw, and Peter Mulder

Center for Chemistry and the Environment,

Gorlaeus Laboratories, Leiden University,

P.O. Box 9502, 2300 RA Leiden, the Netherlands

INTRODUCTION.

Our investigation of the gas phase autoxidation of arenes revealed that above 1000 K
the oxidative degradation proceeds in part by interaction of oxygen atoms with these
compounds [1].

The first step, addition of O(^{3}P) to the aromatic nucleus of e.g. benzene (1), is
well documented [2,3]. The triplet biradical(I) formed, reacts further through interconversion and rearrangement to produce phenol (2) or decomposes to yield a phenoxy
radical (3). The ratio of these reaction channels will depend upon reaction conditions
like pressure, temperatures and concentrations of reactants.

Recently we started to explore these reaction pathways using monosubstituted benzenes
(1) with emphasis on phenol formation.

$$C_6H_5Z + O(^3P) \xrightarrow{(1)} \quad I \quad \begin{array}{l} \xrightarrow{(2)} ZC_6H_4OH \\ \xrightarrow[(-H)]{(3)} ZC_6H_4O\cdot \end{array}$$

EXPERIMENTAL.

Gas phase experiments were performed in a flow system at 80-250 torr and 298 K.
Substrate was introduced by passing a calibrated He stream through an impinger
containing 1. Oxygen atoms were generated by microwave discharge of nitrous oxide
(1% N$_2$O in He). Products were collected in a cold trap and analyzed with GC and/or
GC-MS.

RESULTS AND DISCUSSION.

Phenol and (1,3/1,4) cyclohexadien are the major products observed with $1 = C_6H_6$.
Cyclohexadienes are likely to arise from disproportionation of cyclohexadienyl
radicals, formed from addition of H atoms, generated in (3), to benzene. With
C_6H_6/C_6D_6 mixtures, MS-analyses of the resulting phenols showed that both substrates
led to the respective phenol in equal rate i.e. $k_H/k_D \cong 1$. Isotope analyses disclosed
the presence of C_6H_4DOH and C_6HD_4OH, this H/D incorporation substantiates the
operation of reaction (3). Phenols are partly formed via disproportionation of a

H. Fischer, H. Heimgartner (Eds.)
Organic Free Radicals
© Springer-Verlag Berlin Heidelberg 1988

phenoxy and cyclohexadienyl radical. According to relative ion intensity ca. 30% of the final phenol, under our conditions, is produced via reaction 1,2. Using C_6H_6Z, the ipso product C_6H_5OH is also observed. Rates of phenol formation relative to benzene and isomer distributions are given in the Table.

Z	k_1(rel)	C_6H_5OH/ZC_6H_4OH	ZC_6H_4OH
			o/m/p
CH_3	2.96	0.15	66/22/12
H	(1)	–	–
F	0.57	0.05	47/32/21
Cl	0.82	0.25	49/32/19
CF_3	0.33	0.06	26/58/16

These data show the usual trend for homolytic substitution governed by the addition step (1). Relative rates clearly demonstrates the electrophilic nature of $O(^3P)$ atoms as depicted in the Hammett plot. Interestingly with Z=CF_3 a number of by-products were observed suggesting transfer of F atoms.

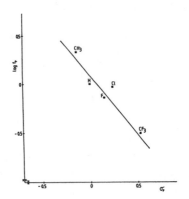

Figure.
Partial rate factors ($\log f_p$) as a function of σ_p.

REFERENCES.

1 Mulder P, Louw R (1986) J Chem Soc Perkin Trans 2 1541
2 Grovenstein E, Mosher AJ (1970) J Am Chem Soc 92:3810
3 Nicovich JM, Gump CA, Ravishankara AR (1982) J Phys Chem 86:1684

ANNULATIONS BY REACTION OF IMIDOYL
RADICALS WITH UNSATURATED BONDS

R. Leardini, D. Nanni, A Tundo and G. Zanardi

Dipartimento di Chimica Organica
Università di Bologna
Viale Risorgimento, 4 - 40136 Bologna, Italy

We have been studying for a few years the reactivity of arylimidoyl radicals generated by hydrogen abstraction from Schiff's bases by means of peroxides; the first useful reaction was the synthesis of quinoline derivatives, performed by reacting imines with mono-substituted alkynes in benzene at 60°C in the presence of di-isopropylperoxydicarbonate (DPDC) (1). In a subsequent paper we found the mechanism of this reaction to proceed also _via_ an intermediate spirocyclohexadienyl radical (2); the reaction scheme may be written as follows:

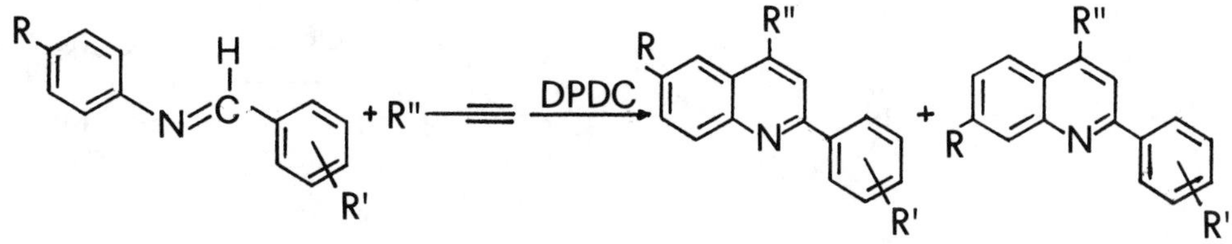

When the same reaction is carried out with alkenes, we still obtain quinolines (in lower yields with respect to the reaction with alkynes) but the reaction does not seem to involve the spiro-radical:

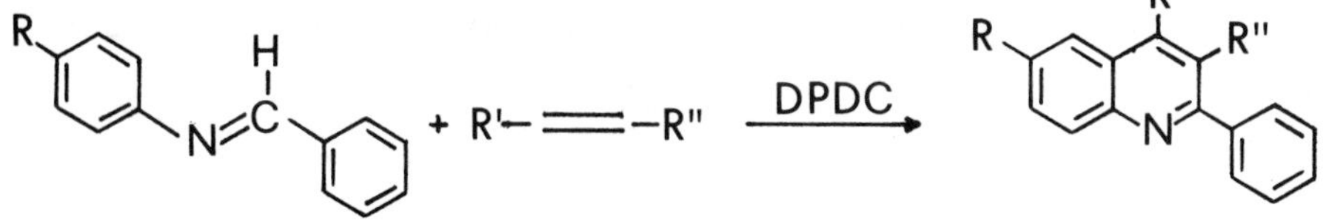

R = Cl, Me, OMe, COMe
R′ = Ph, CN, COOMe, SiMe$_3$
R″ = H, CN, COOMe

H. Fischer, H. Heimgartner (Eds.)
Organic Free Radicals
© Springer-Verlag Berlin Heidelberg 1988

The reaction with alkynes may follows a different pathway when we change both temperature and peroxide; in fact, N-benzylideneaniline reacts with diphenylacetylene and DPDC at 60 °C in benzene to afford 2,3,4-triphenylquinoline, but it gives 2,3-diphenylindenone as main product, together with 2,3,4-triphenylquinoline, when allowed to react with the same alkyne in chlorobenzene at 115 °C in the presence of di-t-butylperoxide (TBP):

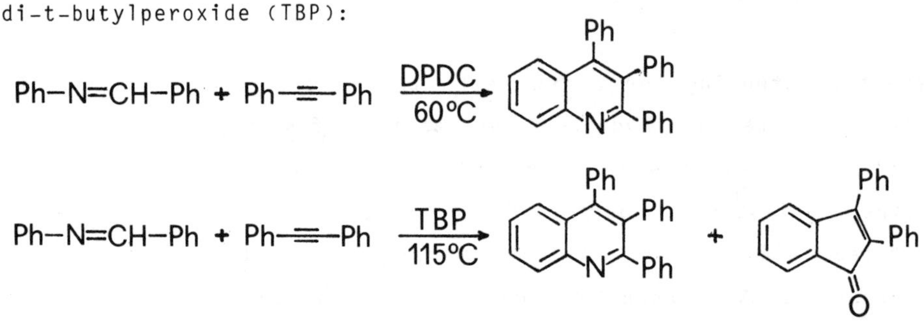

Ph—N=CH—Ph + Ph—≡—Ph $\xrightarrow[60°C]{DPDC}$ [2,3,4-triphenylquinoline structure]

Ph—N=CH—Ph + Ph—≡—Ph $\xrightarrow[115°C]{TBP}$ [2,3,4-triphenylquinoline structure] + [2,3-diphenylindenone structure]

A very interesting reaction is the annulation performed with imines and diethylazodicarboxylate under usual conditions:

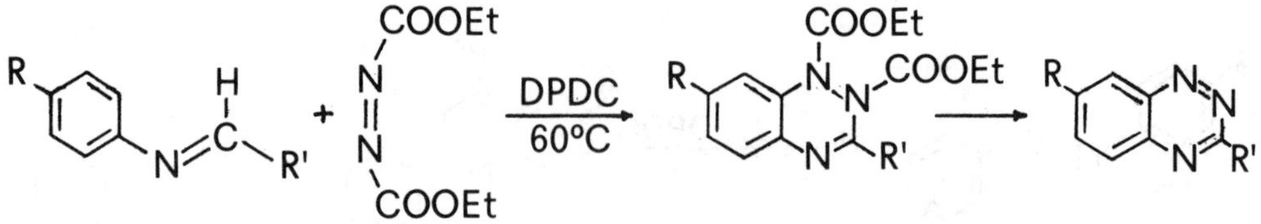

[reaction scheme with imine + azodicarboxylate → DPDC 60°C → intermediate → benzotriazine products]

R = H, OMe
R´ = Ph, t-Bu
Yields = 27-72%

The reaction represents a new example of intermolecular radical addition to the azo group and a novel synthetical route to benzo-
-1,2,3-triazines and their 1,2-dihydro derivatives.

References

1) R. Leardini, G.F. Pedulli, A. Tundo, and G. Zanardi (1984) J.Chem.Soc., Chem.Commun., 1320.
2) R. Leardini, D. Nanni, G.F. Pedulli, A. Tundo, and G. Zanardi (1986) J.Chem.Soc., Perkin Trans. 1, 1591

SELF ELECTRON TRANSFER REACTIONS
OF HYDRAZINES

S.F. Nelsen

Department of Chemistry, University of Wisconsin
Madison, Wisconsin 53711

SELF ELECTRON TRANSFER AND ΔHr

Self electron transfer (ET) is used here to mean electron exchange between the neutral form (**n**) and the cation radical (**c**) of the same compound. Organic compounds with multi-atom pi systems such as tetramethyl-p-phenylene-diamine show second order rate constants for self ET (k_{et}) above 10^8 $M^{-1}s^{-1}$ in acetonitrile solution. In constrast, most tetraalkylhydrazines, which have two electron pi systems, show much slower self-ET, with rate constants below about 10^3 $M^{-1}s^{-1}$ [1]. Because self ET has $\Delta G° = 0$ and is always at equilibrium under ordinary conditions, magnetic resonance line broadening is used to measure k_{et}, and virtually no broadening can be observed for most tetraalkylhydrazines. Their slow ET results from the unusually large geometry change which occurs upon removal of an electron, resulting in large relaxation enthalpies, which are quantitatively the difference between vertical and adiabatic ionization potentials: ΔHr = vIP -aIP. It has proven possible to measure ΔHr for several examples using High Pressure Mass Spectrometry [2], and the influence of alkyl group size and shape on tetraalkylhydrazine ΔHr is rather well understood. Because ΔHr is close to half the Marcus "inner shell" reorganization energy, tetraalkylhydrazines should be the first class of organic compounds studied for which the major part of the observed barrier is caused by internal reorganization.

SESQUIBICYCLIC HYDRAZINES AND THEIR CATION RADICALS

The addition of protonated bicyclic azo compounds to cyclic dienes [3,4] makes N,N'-bis-bicyclic hydrazines, "sesqubicyclic" ones, easily available. The special geometrical features of the alkyl groups allow isolation of the cation radicals for several sesquibicyclic tetraalkylhydrazines. The amount of geometry reorganization which occurs upon electron loss has been documented for sesquibicyclic hydrazines by determination of x-ray crystallographic structures for both **n** and **c** [5]. Semiempirical MO calculations do a surprisingly good job at calculating the change in pyramidality at nitrogen upon electron loss for these compounds.

k_{et} VALUES FOR SESQUIBICYCLIC HYDRAZINES

Our principal reason for interest in sesquibicyclic hydrazines has been that they show small enough geometry reorganization that their k_{et} values are small enough to show significant NMR line

H. Fischer, H. Heimgartner (Eds.)
Organic Free Radicals
© Springer-Verlag Berlin Heidelberg 1988

broadening, allowing quantitative measurement of k_{et}. A preliminary study of k_{et} values for the 2,3-diazabicyclo[2.2.2]oct-2-ene, cyclohexadiene adduct have already been reported [6], as has the use of semiempirical MO calculations for consideration of the ET barrier observed [7]. We shall focus here on solvent effects on k_{et} for this compound, which are anomalous. Activation parameters for self ET of sesquibicyclic hydrazines will be compared with those for tetramethyl-p-phenylene-diamine, and the small size of k_{et} shown to result from far larger "inner shell" reorganization energy for the hydrazines. In contrast to previously studied compounds, linear plots of $\ln(k_{et})$ with the Marcus solvent parameter, gamma, are not observed for hydrazines. Nine solvents, including three alcohols, do show linear $\ln(k_{et})$ versus Kosower Z value or Dimroth E_T value. Both the Kosower and Dimroth solvent parameters are direct experimental measurements of the optical Marcus ET barrier, lambda, for specific compounds. Optical Marcus ET barriers from several other compounds have been found to give linear plots with the Marcus solvent parameter, gamma. The principal empirical difference seems to be the size of the Marcus lambda value. Solvation energies of hydrazine cation radicals will be shown to correlate with gas phase alkyl group polarizability, indicating that it is not the formally half-positive nitrogens, but the alkyl groups that show the principal interation with the solvent in these cation radicals.

References:
1 For a review see: Nelsen SF (1986) In: Liebman JF, Greenberg A (eds) Molecular Structures and Energetics. VCH, Deerfield Beach FL, Vol 3
2 Meot-Ner(Mautner) M, Nelsen SF, Willi MR, Frigo TB (1984) J Am Chem Soc 106:7384
3 Nelsen SF, Blackstock SC, Frigo TB (1984) J Am Chem Soc 106: 3366.
4 Nelsen SF, Blackstock SC, Frigo TB (1986) Tetrahedron 42:1769
5 Nelsen SF, Blackstock SC, Haller KJ (1986) Tetrahedron 42:6101
6 Nelsen SF, Blackstock SC (1985) J Am Chem Soc 107:7189
7 Nelsen SF, Blackstock SC, Kim Y (1987) J Am Chem Soc 109:677

A Carbon-Centred Alkylphosphonate Radical
Generated by Pyruvate Formate-Lyase with its Substrate Analogue Hypophosphite

Volker Unkrig*, Joachim Knappe*, and Franz A.Neugebauer[+]

*Institut für Biologische Chemie, Universität Heidelberg, and
[+]MPI für medizinische Forschung, Abt. Organische Chemie, D-6900 Heidelberg, FRG

Abstract: Pyruvate formate-lyase, a homodimeric protein of 2 x 85 kDa, is distinguished by the property of containing a stable organic free radical ($g = 2.0037$) in its structure. The enzyme catalyzes pyruvate conversion to acetyl-CoA via two distinct half-reactions (E-SH + pyruvate $\rightleftharpoons$ E-S-acetyl + formate; E-S-acetyl + CoA $\rightleftharpoons$ E-SH + acetyl-CoA), the first of which has been proposed to occur through reversible carbon-carbon bond cleavage by a homolytic mechanism (1).

For the enzyme-based radical (R$^{\bullet}$) on E-SH as well as on E-S-acetyl, the unpaired electron was found to be coupled to various hydrogen nuclei [$a = 1.5$ mT (1H); $a = 0.37$ mT (?H) and $a = 0.59$ mT (?H)]; the one giving rise to the principal doublet splitting of the EPR spectrum representing an acidic proton that exchanges in D_2O. The chemical structure, probably a modified amino acid residue, has not yet been identified. [The possibility of sulfur radical has been excluded via ^{33}S (I = 3/2) enriched enzyme samples].

When the formate-analogue hypophosphite is added to E-S-acetyl at 0 °C a novel radical species, which gradually accumulates, was found. Its multiplet-structured EPR spectrum is centered at $g = 2.0032$ and shows hyperfine couplings to 1 P and 3 H [$a = 2.72$ mT (1P) and $a = 1.96$ mT (3H)]. These assignments are based on experiments with deuterated substrates and deuterated enzyme. The EPR data are consistent with the structure of an α-phosphonyl carbon-centred radical (1B) as they compare well with that of the previously characterized radical H_3C-C$^{\bullet}$(OH)-P(O)(OCH$_3$)$_2$ (2).

Although stable at 0 °C (T/2 > 20 min), the accumulated radical is quickly quenched when the solution is brought to ambient temperature. Protein chemical analyses (3) have revealed that the enzyme after this treatment contains 1-hydroxy-ethyl-phosphonate in thioester linkage to cysteinyl residue 418 (1C). This amino acid residue is adjacent to Cys-419 which functions as the covalent-catalytic acetyl carrier (1A).

H. Fischer, H. Heimgartner (Eds.)
Organic Free Radicals
© Springer-Verlag Berlin Heidelberg 1988

The overall process

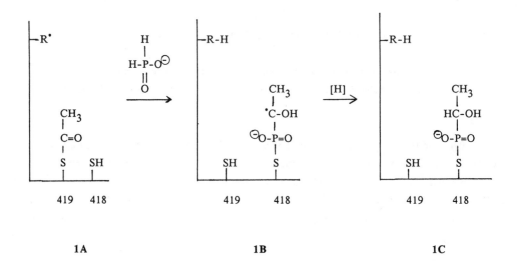

1A   1B   1C

may be initiated by H-abstraction from hypophospite by the enzyme based radical (R˙). We propose that the demonstrated formation of **1B** mimics a putative radical intermediate of the physiological process when formate reacts with E-S-acetyl to produce pyruvate (reversed direction of the first half reaction, see above).

1 Knappe J, Neugebauer FA, Blaschkowski HP, Gänzler M (1984) Proc Natl Acad Sci USA 81:1332
2 Damerau W, Laßmann G (1971) Z Chem 11:182
3 Plaga W, Frank R, Knappe J unpublished

Radical Ions of Benzo[c]cinnolines and 2,3-Dihydro-1H-benzo[c]pyrazolo[1,2-a]cinnolines

H. Chandra[+], H. Fischer[*], F.A. Neugebauer[*] and M.C.R. Symons[+]

[+]Department of Physical Chemistry, The University, Leicester LE1 7RH, England, and
[*]MPI für medizinische Forschung, Abt. Organische Chemie, D-6900 Heidelberg, FRG

Abstract: The radical cations of benzo[c]cinnoline (1) and its four symmetrical dimethyl derivatives have been prepared (CFCl$_3$, ^{60}Co γ-irradiation, 77 K). The e.s.r. spectra were all characterized by large hyperfine coupling to two equivalent nitrogen atoms. Analysis of the data gave ca. 9 % 2s and 44 % p character on each nitrogen, thereby establishing a σ-structure. Similar treatment of the 2,9- and 3,8-dimethoxy derivatives gave a broad unresolved singlet. This shows that the SOMO has switched from a σ(N)-orbital to a π-orbital.

Exposure of dilute solutions of all six compounds as dilute solutions in CD$_3$OD gave parallel features characteristic of two equivalent ^{14}N nuclei. The perpendicular splittings were close to zero, thus confirming the SOMO is π. These anions were also prepared in DME and HMPT, and their e.s.r. and endor spectra were recorded. This led to full assignments in all cases. Our assignment for the anion of the parent compound differs from that previously proposed.

In an extension of previous work on 5,6-dihydrobenzo[c]cinnolines radical cations (1) we also studied the radical cations derived from 2,3-dihydro-1H-benzo[c]pyrazolo[1,2-a]cinnolines (2).

1 Neugebauer FA, Bock M, Kuhnhäuser S, Kurreck H (1986) Chem Ber 119:980

H. Fischer, H. Heimgartner (Eds.)
Organic Free Radicals
© Springer-Verlag Berlin Heidelberg 1988

AN ELECTRON SPIN RESONANCE STUDY
OF SOME SILANONE RADICAL ANIONS

Alwyn G. Davies and Anthony G. Neville

Chemistry Department, University College London,
20 Gordon Street, London, WC1H OAJ, U.K.

A method has been developed for generating silanone radical anions
from simple precursors. The t-butoxyl radical will abstract a
hydrogen atom from a hydroxysilane (1) to give the hydroxysilyl
radical (2). Treatment of (1) with t-butoxide solutions yields the
siloxide (3) from which hydrogen atom abstraction gives the desired
silanone radical anion (4).

Thus, both the di-t-butylhydroxysilyl radical (2a, R^1, R^2 = Bu^t)
and the di-t-butylsilanone radical anion (4a, R^1, R^2 = Bu^t) have
been generated[1]. The e.s.r. spectra of both species are well
resolved multiplets which show many notable differences.

Hyperfine coupling to the hydroxylic hydrogen is clearly
observed in the spectrum of (2a) which contrasts with its absence in
the spectrum of (4a), where coupling to a sodium counter ion is
apparent. Also accompanying this change is a decrease in the
coupling to the 18 equivalent t-butyl hydrogens, an increase in the
g-value and a significant decrease in the silicon coupling. The
change in silicon coupling is consistent with a more planar radical

H. Fischer, H. Heimgartner (Eds.)
Organic Free Radicals
© Springer-Verlag Berlin Heidelberg 1988

centre in the silanone radical anion, which is also indicated by MNDO and *ab initio* calculations.

Whilst less hindered hydroxysilanes are not readily isolated, further related species have been generated from the severely hindered hydroxysilanes, (5), including the sole silanal anion yet observed (6a, R = H) and sila-acylfluoride radical anion (6f, R = F). Where observed, each of these 6 conjugate acid radical / radical anion pairs show effects that parallel the above example.

$$\text{TsiSiR(OH)H} \qquad\qquad \begin{array}{c} \text{Tsi} \\ \diagdown \\ \text{Si}\overset{\cdot}{=}\text{O} \\ \diagup \\ \text{R} \end{array}$$

$$\qquad\quad \textbf{(5)} \qquad\qquad\qquad\qquad \textbf{(6)}$$

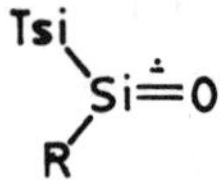

$$\text{Tsi} = (\text{Me}_3\text{Si})_3\text{C} \qquad \text{R} = \text{H, Me, Et, Bu}^n\text{, Ph, F}$$

The work is being extended to smaller symmetrical silanone radical anions by direct synthesis of the siloxides (3), and by analogous methods to generate radical anions showing partial multiple bonding between silicon and sulphur and silicon and nitrogen.

REFERENCE

(1)   A.G. Davies and A.G. Neville, J. Chem. Soc., Chem. Commun., 1987, 16

NUCLEAR-ELECTRON RESONANCE IN ^{31}P-CONTAINING FREE RADICAL SOLUTION

B.M. Odintsov, A.V. Il'yasov and R.G. Yahin

Kazan Branch USSR Academy of Sciences
Lobachevsky Str., 2/31 Kazan 420111

Abstract: Dynamic nuclear polarization (DNP) and ESR spectra in the solutions of ^{14}N- and ^{31}P-containing free radicals in the intermediate magnetic fields were investigated by double nuclear-electron resonance (DNER) method. Qualitatively new dependence of polarization value on paramagnetic centres concentration was revealed.

INTRODUCTION

The decrease of the enhancement coefficient of DNP under the influence of electron spin exchange in the paramagnetic solutions with the single ESR line is well known. At the same time effect of anomalous DNP increase due to the exchange interactions in the solutions of paramagnetic particles with the resolved hyperfine structure (HFS) is practically not investigated.

Method

Double nuclear-electron resonance is based on the DNP phenomenon. Up to now DNER is used as an effective method of investigation of hyperfine interactions and dynamic processes in paramagnetic solutions. This method is widely applied, especially when other methods of detecting very weak intermolecular shorttime coupling such as a direct observation of hyperfine structure in ESR and shift measurements of NMR are uneffective due to line broadening. DNER appeared to be the most sensitive method of ESR spectra registration in weak magnetic fields /1/.

Results and discussion: In intermediate magnetic fields there was made the investigation on dynamic polarization of hydrogen nuclei of solvent molecules in the nitrogen- and phosphorus-containing radicals remaining well resolved HFS in ESR spectrum. For all inves-

H. Fischer, H. Heimgartner (Eds.)
Organic Free Radicals
© Springer-Verlag Berlin Heidelberg 1988

tigated compounds large negative polarization testifying to the di-
pole-dipole type of interconnection between magnetic moments of radical molecules and protons of solvent molecules was observed.   In
the representation of complete angular moment $|F,m_F\rangle$ of a molecule
the calculation of energy scheme and peaks positions of ESR-absorb-
tion of the investigated radicals in zero, weak and   intermediate
magnetic fields with regard to the hyperfine splitting on the phosphorus and nitrogen nuclei was carried out. Owing to the absence of
the analytical solution of the problem, the program of the numerical
calculation of the dynamical enhancement coefficients of nuclei polarization for the particles with some magnetononequivalent nuclei
was made according to expression:

$$A = \gamma_S/\gamma_I \cdot \xi f \, (SNS_o)^{-1} \sum_{F,m_F} < F,m_F|S_z|F,m_F> n_m \; ,$$

which take into account the redistribution of spin populations in a
whole system of energy levels at the saturation of individual ESR-
transitions. Numerical calculation results show that the value and
sign of dynamical nuclei polarization depends on the saturated ESR-
transitions. Besides it was discovered that as the concentration of
paramagnetic is increased the experimental values of DNP enhancement noticeably exceeds ones predicted by the theory for dipole-di-
pole interaction. Numerical calculation indicates that the experimental results are two or more times as much as the theoretically
predicted ones. The detailed analyses point out that the only reason for additional DNP enhancement is the exchange interaction   of
electron spins in free radical solutions. In DNP there are only   a
few number of theoretical works taking into account the exchange
interactions for paramagnetic solutions with the resolved HFI. The
obtained theoretical expressions decribe qualitatively well an additional enhancement of polarization which is due to the additional
redistribution of populations in the system of electron levels. It
is necessary to stress that an additional DNP enhancement is increased with the growth of the paramagnetic concentration and of the
exchange frequency, but the hyperfine components, whose saturation
leads to great polarization, being well resolved in the ESR spectrum. In this respect the radicals containing phosphorus can be particularly interesting, since they have great hyperfine splitting of
the nucleus ^{31}P preventing HFS convolution in the concentrated solutions into one exchanging line, when polarization abrupty falls.

1 Odintsov BM (1986) Electron-nuclear Overhauser effect in solutions. Kazan Branch USSR Academy of Sciences Press, Kazan

NITRO COMPOUNDS AS FREE RADICAL PRECURSORS

N. Ono

Department of Chemistry, Faculty of Science,
Kyoto University, Kyoto 606, Japan

Abstract: Treatment of nitro compounds with tin radicals affords alkyl radicals, which undergo radical cyclization, addition to electron deficient alkenes, and β-elimination reactions. Combination of these reactions with carbon-carbon bond forming reactions of nitro compounds provides a useful synthetic method.

Aliphatic nitro groups except for primary ones are readily replaced by hydrogen on treatment with Bu_3SnH in the presence of AIBN at 80-110 °C.[1]

$$R-NO_2 \;+\; Bu_3SnH \;\xrightarrow{\text{AIBN}}\; R-H \;+\; Bu_3SnONO$$

R = Tertiary, benzyl yield 80-90 %

R = Secondary   yield 40-60%

Based on this denitration, following new synthtic methods are devised.

1. Conjugate Addition of Alkyl groups.[2] The Michael addition of nitro compounds and sunsequent removal of the nitro group from the adduct provides a useful method for the conjugate addition of alkyl groups.

2. Synthesis of Ketones.[3]

3. Regioselective Diels-Alder Reactions. The sequence of the Diels-Alder reaction of nitroalkenes with dienes and subsequent denitration provides a new method for the regioselective construction of cyclohexene derivatives.[4]

4. Synthesis of Deuterated Compounds. Nitro groups are selectively replaced by deuteride on treatment with Bu_3SnD, which provides a new method for the preparation of deuterated compounds.[5]

H. Fischer, H. Heimgartner (Eds.)
Organic Free Radicals
© Springer-Verlag Berlin Heidelberg 1988

5. Aliphatic Nitro Compounds as Radical Precursors. As reductive cleavage of aliphatic nitro compounds with Bu_3SnH proceeds via radical intermediates, nitro compounds are used as precursors to alkyl radicals.[6] For example, secondary nitro compounds are doubly alkylated by two kinds of electron deficient alkenes.

$$R^2-\underset{NO_2}{\underset{|}{\overset{R^1}{\overset{|}{C}}}}-H \xrightarrow[\text{base}]{CH_2=CH-Y} R^2-\underset{NO_2}{\underset{|}{\overset{R^1}{\overset{|}{C}}}}-CH_2CH_2-Y \xrightarrow[\substack{Bu_3SnH \\ AIBN}]{CH_2=CH-Z} R^2-\underset{CH_2CH_2Z}{\underset{|}{\overset{R^1}{\overset{|}{C}}}}-CH_2CH_2Y$$

$\underline{1}$

R^1	R^2	Y	Z	Yield of $\underline{1}$, %
CH_3	CH_3	$CO_2C_2H_5$	CO_2CH_3	60
CH_3	$C_6H_5CH_2$	CN	CO_2CH_3	54
CH_3	$C_6H_5CH_2$	CN	CN	44
CH_3	CH_3	$COCH_3$	CO_2CH_3	60

6. Olefin Synthesis. Reductive cleavage of nitro groups in compounds $\underline{2}$ having radical leaving groups to the nitro group gives olefin.[7]

$$R^2-\underset{NO_2}{\underset{|}{\overset{R^1}{\overset{|}{C}}}}-\underset{X}{\underset{|}{\overset{R^3}{\overset{|}{C}}}}-R^4 \quad + \quad Bu_3SnH \longrightarrow \underset{R^2}{\overset{R^1}{\diagdown}}C=C\underset{R^4}{\overset{R^3}{\diagup}}$$

$\underline{2}$

The elimination reaction proceeds with high stereospecificity.[8] Thus, (E)-Allyl alcohols are prepared by the reaction of anti-β-nitro sulfides ($\underline{3}$), which are obtained from nitroolefins, thiophenol, and formaldehyde.

$$R-CH=CH-NO_2 \quad + \quad PhSLi \quad + \quad HCHO \longrightarrow R-\underset{SPh}{\underset{|}{\overset{NO_2}{\overset{|}{CH}}}}-CH-CH_2OH \xrightarrow[AIBN]{Bu_3SnH}$$

$\underline{3}$

$$R-CH=CH-CH_2OH$$

$\underline{4}$

1)N. Ono and A. Kaji, Synthesis, 1986, 693; 2). N. Ono, et al J. Org. Chem., 50, 3692 (1985); 3) N. Ono, et al. Synthesis, 1987, 532; 4)N. Ono, et al J. Chem. Soc. Perkin I, 1987, 1929; 5)N. Ono, et.al Chem. Lett., 1982, 1079; 6)N. Ono, et.al Tetrahedron, 1985, 4013; 7) N. Ono, et al Chem. Lett., 1981, 1139; 8) N. Ono, et al J. Org. Chem., 52, 5111 (1987).

FREE RADICALS IN THE CATALYTIC
CHEMISTRY OF HEMOPROTEINS

Paul R. Ortiz de Montellano

Department of Pharmaceutical Chemistry,
School of Pharmacy, University of California
San Francisco, California 94143-0446, U.S.A.

The monooxygenases and peroxidases, two of the major types of catalytic hemoproteins, are catalytically distinguished in that cytochrome P-450 enzymes transfer an oxygen atom to their substrates whereas peroxidases abstract one electron from theirs. In general, diffusible free radicals are produced by the peroxidases but not monooxygenases. Solid evidence nevertheless suggests that cytochrome P450-catalyzed reactions proceed via free radical intermediates that are trapped before they diffuse away from the enzymatically generated ferryl oxygen species.

The differential reactions of horseradish peroxidase and cytochrome P450 with alkylhydrazines, arylhydrazines, and sodium azide provide evidence for key structural differences in the active sites of these enzymes. Oxidation of the hydrazines by cytochrome P450 monooxygenases produces radicals that react with the iron and nitrogens of the prosthetic heme group, whereas oxidation of both the hydrazines and the azide anion by the peroxidases produces radicals that react exclusively with the delta-meso carbon of the heme group. These results suggest that substrates interact exclusively with the heme edge in the peroxidases but with the ferryl oxygen in the monooxygenases.

Free radicals do escape into the medium in the cytochrome P450-catalyzed oxidations of alkylhydrazines and 4-alkyl-1,4-dihydropyridines. The dihydropyridines are oxidized by cytochrome P450 to dihydropyridine radical cations that aromatize, in part, by extruding the 4-alkyl group as a free radical. Recent work indicates that cytochrome P450 also catalyzes a net dehydrogenation of the dihydropyridines by a related free radical pathway. Detailed studies of the oxidation of dihydropyridine probes by cytochrome P450, including parallel studies of their chemical oxidation by ferricyanide, shed substantial light on the radical intermediates formed in nitrogen oxidation reactions.

Radicals appear to escape from the catalytic sites of both the peroxidases and cytochrome P450 when the ferryl oxygen is physically separated from the radical species. This is achieved in the peroxidases by the active site architecture but in cytochrome P450 enzymes probably only occurs when fragmentation of an initial transient substrate radical unmasks a second free radical some distance away from the ferryl oxygen.

Acknowledgements: Support of this work by the National Institutes of Health is gratefully acknowledged.

1. Ortiz de Montellano, PR (1987) Accounts Chemical Research 20:289
2. Ator, MA, Ortiz de Montellano, PR (1987) J. Biol. Chem. 262:1542
3. Augusto, O., Beilan, HS, Ortiz de Montellano, PR (1982) J. Biol. Chem. 257:11288
4. Lee, JA, Jacobsen, NE, Ortiz de Montellano, PR, Biochemistry, submitted for publication.

H. Fischer, H. Heimgartner (Eds.)
Organic Free Radicals
© Springer-Verlag Berlin Heidelberg 1988

SPIN TRAPPING WITH THIONES

A. ALBERTI,[a] M. BENAGLIA,[a] B. F. BONINI[b] and G. F. PEDULLI[b]

[a]I.Co.C.E.A. - C.N.R., I-40064 OZZANO EMILIA, Italy
[b]Dipartimento di Chimica Organica, Università, I-40127 BOLOGNA, Italy

Thiones are known to be efficient spin traps for a variety of free radicals centred at carbon and at other elements of Group IV (silicon, germanium, tin, lead), Group V (phosphorus), Group VI (oxygen, sulphur, selenium) and at transition metals (such as manganese and rhenium). The use of thiones as spin traps is however limited by the rather low stability of these compounds, which must generally be used at low temperatures and under inert atmospheres.

The synthesis of some relatively stable thiocarbonyl compounds have been reported recently, namely that of thiobenzoyltriphenylsilane [**TBTPS**, $Ph_3SiC(S)Ph$] by Bonini et al. and those of tris(trimethylsilyl)ethanethial (or trisylthioaldehyde) [**TSTA**, $(Me_3Si)_3CC(S)H$] and 2,4,6-tri-tert-butylthiobenzaldehyde [**TBTBA**] by Okazaki et al.

We have found that these three thiocarbonyl compounds act as versatile spin traps and react with a wide number of organic radicals to give paramagnetic adducts which can be investigated by ESR spectroscopy. In particular, **TBTPS**, the most versatile of the three substrates, leads to very persistent spin-adducts, whose spectra allow in most cases an unambiguous identification of the trapped radicals, and is particularly convenient for trapping species centred at Group IVB elements or at transition metal atoms, as well as alkoxyl and thiyl or sulphonyl radicals. The remarkable reactivity of thiobenzoyltriphenylsilane is reflected in the detection of the $Ph(Ph_3Si)\dot{C}-SH$ adduct, which represents the first observation of a radical with the SH function bonded to the radical carbon, a species often invoked in

H. Fischer, H. Heimgartner (Eds.)
Organic Free Radicals
© Springer-Verlag Berlin Heidelberg 1988

the photochemistry of thiones, but as yet elusive in solution.

Although also the two thioaldehydes react readily with a large number of free radicals, their adducts show in all cases very similar ESR spectra, which are little informative about the nature of the trapped species. Therefore, also in view of their tedious synthetic procedures, their use as spin traps does not show any particular advantages over that of other commercial or more readily synthesizable trapping agents.

By making use of competitive techniques, the rates of addition of *tert*-butoxyl and pivaloyl radicals to **TSTA** were measured as $(1.2\pm0.5)\times10^7$ and $(1.9\pm0.5)\times10^6$ $M^{-1}s^{-1}$, respectively. These values indicate that radical trapping by thiocarbonyls is a fast process, probably owing to the low strength of the π C-S bond, and are consistent with the results of previous studies reported in the literature.

INTRAMOLECULAR MOTION IN ORGANIC FREE RADICALS STUDIED BY MUON SPIN ROTATION AND LEVEL-CROSSING SPECTROSCOPY

P.W. Percival[1], J.-C. Brodovitch[1], S.-K. Leung[1], D. Yu[1]
R.F. Kiefl[2], D.M. Garner[2], G.M. Luke[2], K. Venkateswaran[2]
and S.F.J. Cox[3]

[1] Chemistry Department, Simon Fraser University,
Burnaby, B.C., Canada V5A 1S6
[2] TRIUMF, Vancouver, British Columbia, Canada V6T 2A3
[3] RAL, Chilton, Didcot, Oxfordshire OX11 0QX, U.K.

Abstract: Muon spin rotation and level-crossing spectroscopy have been used to study intramolecular motion in several muonium- substituted free radicals, including ethyl, tert-butyl, and cyclohexadienyl.

INTRODUCTION

Muonium-substituted radicals (1) are formed when muonium adds to alkenes, e.g. the ethyl radical $MuCH_2CH_2$ is formed when ethene is irradiated with muons. The muonium atom may be considered as a light ($m=m_H/9$) isotope of hydrogen, and the hyperfine interaction between the muon (a spin-½ particle) and the unpaired electron is governed by the same factors as the proton hyperfine coupling constant (hfc) in the corresponding hydrogen radical. The muon hfc can be measured by muon spin rotation spectroscopy (μSR), and analysis of its temperature dependence provides information about the configuration and conformation of radicals in analogous manner to conventional ESR studies (2). With the recent development of muon level-crossing spectroscopy (μLCR) it is now possible to determine other (i.e. non muon) nuclear hfcs in a muonium-substituted radical (3). As well as providing a very significant increase in the information on radical structure and dynamics, μLCR is remarkably versatile, in that it can be applied to the study of radicals in the solid, liquid, solution and gaseous states. In the following, we briefly summarize our studies of intramolecular motion in tert-butyl, ethyl and cyclohexadienyl.

H. Fischer, H. Heimgartner (Eds.)
Organic Free Radicals
© Springer-Verlag Berlin Heidelberg 1988

tert-BUTYL (4)

Muon (A_μ) and proton (A_p) hfcs for the tert-butyl radical $(CH_3)_2CCH_2Mu$ have been determined over a wide range of temperature in pure iso-butene. $A_p(CH_3)$ is almost constant, but $A_\mu(CH_2Mu)$ falls and $A_p(CH_2Mu)$ rises with increasing temperature, consistent with a preferred con-formation of the methyl group in which the C-Mu bond is coplanar with the symmetry axis of the radical orbital. The A_μ data covers the temperature range from 297K down to 43K, where the solution is frozen. There is a discontinuity in A_μ at the melting point, as well as a change in temperature dependence. We suggest that the potential barrier for methyl group rotation is lower in the liquid due to simul-taneous inversion at the radical centre, and that the inversion mode is somewhat inhibited in the solid. A simultaneous fit of A_μ and A_p indicates a V_2 barrier of 1.8 kJ/mol, and is consistent with a long C-Mu bond and a tilt of the CH_2Mu group in the direction that brings the Mu atom closer to the radical centre.

ETHYL

Unlike tert-butyl, A_μ in the ethyl radical $MuCH_2CH_2$ varies smoothly from 20 K in solid ethene, through the liquid range, to 300 K in the gas. Preliminary analysis suggests a torsional barrier of 3.1 kJ/mol.

CYCLOHEXADIENYL

A negative temperature-dependence was found for the muon and proton hfcs in the methylene groups of C_6H_6Mu, C_6D_6Mu and C_6F_6Mu, but the ^{13}C hfc was constant. The results are consistent with bending vibra-tions of the C-Mu and C-X bonds without distortion of the planar ring configuration.

References:

1   Roduner E (1984) In: Chappert J, Grynszpan (eds) Muons and pions in materials research. North-Holland, Amsterdam.
2   Kochi J K (1975) In: Williams G H (ed) Advances in free radical chemistry, volume V. Academic Press, New York.
3   Percival P W, Kiefl R F, Kreitzman S R, Garner D M, Cox S F J, Luke G M, Brewer J H, Nishiyama K, Venkateswaran K (1987) Chem Phys Letters 133:465
4   Percival P W, Brodovitch J C, Leung S K, Yu D, Kiefl R F, Luke G M, Venkateswaran K, Cox S F J (1988) Chem Phys, submitted

STEREOCHEMICAL AND REGIOCHEMICAL
ASPECTS OF FREE RADICAL MACROCYCLIZATION

Ned A. Porter

Department of Chemistry, Paul M. Gross Chemical Laboratory

Duke University, Durham, NC, 27706, USA

The use of free radicals for the construction of ring systems has proved to be a very powerful tool for the synthetic chemist. Rings having five or six members are readily prepared by free radical methods and the intramolecular addition of carbon radicals to olefins and aldehydes provides ready access to cycloalkanes and cycloalkanols [1-3]. Rate constants for intramolecular addition of carbon radicals to simple olefins are greatest for 5-*exo* cyclizations and these rate constants drop off sharply for rings greater than six members [4-6]. Thus, radical cyclization has for the most part been limited to the preparation of five and six membered rings. We have studied the intramolecular addition of radicals to olefins activated with electron withdrawing substituents such as cyano and carbonyl groups and we find that 14-20 membered rings can be prepared with yields as high as 90%. Thus, the extension of free radical synthetic methodology.to the construction of large rings should prove to have many applications.

Scheme I

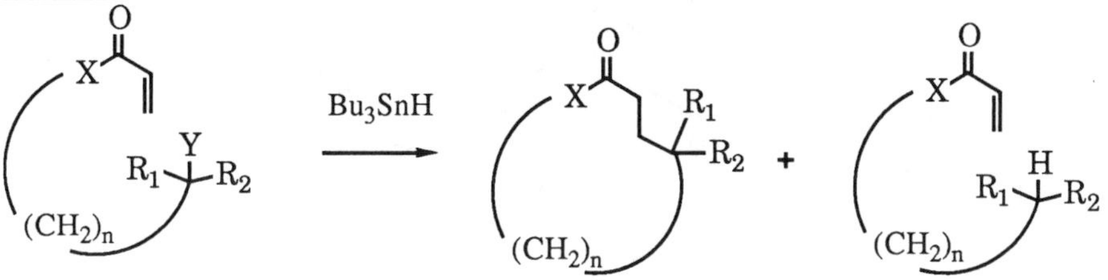

$$1 \qquad\qquad 2 \qquad\qquad 3$$

Several substrates such as **1** have been studied. Yields of macrocycle **2** are typically 50-60% for reactions carried out with substrates having n= 9 to 15, X= CH_2 or O, R_1=R_2=H and Y=I [7,8]. Typically reactions are run in benzene with 0.1 equivalents AIBN initiator at 3-10 mM **1** and 1.1 equivalents of tin hydride. For **1** with R_1 and/or R_2= CH_3, yields of macrocycle are high, 80-90%, since the reactions can be run at low concentrations (0.7 mM) and the chain still propagates. Reactants with Y=Br do not give good yields of macrocycles under the standard tin hydride conditions.

Cyclization reactions carried out on substrates such as **1** always give endocyclic products. If, however, the alkene that is the target for intramolecular radical addition is suitably substituted, exocyclization competes with endocyclization. Thus, compounds with target alkene having the substructures as shown below give both endocyclic and exocyclic products. The ratio of endocyclic

H. Fischer, H. Heimgartner (Eds.)
Organic Free Radicals
© Springer-Verlag Berlin Heidelberg 1988

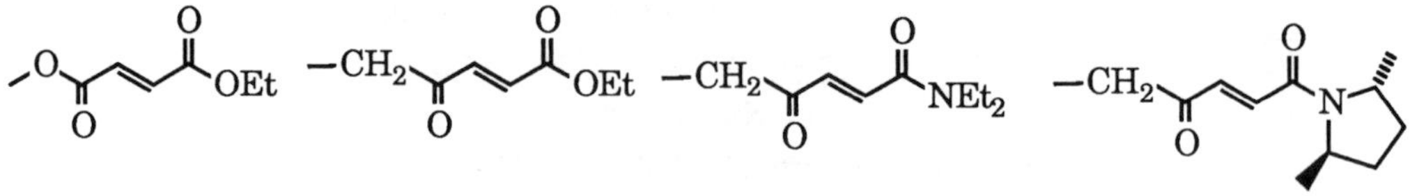

to exocyclic products depends on the ring size and the substitution on the alkene with endocyclic products being generally favored [7,8]. At large ring sizes, however, endocyclic:exocyclic product ratios are nearly 1:1. Only exocyclic products are observed with substitution -CH=C(CN)$_2$ on the alkene [9]. Cyclization to the dimethylpyrollidine amide occurs with significant diastereoselectivity [10].

Macrocyclization-transannular cyclizations can also be achieved in one step. Thus, as shown in Scheme II, yields of bicyclo [9.3.0] systems as high as 50% are observed from an acyclic precursor [9]. Cis and trans-fused products are observed and epimerization by base gives predominantly the trans ring-fused product .

Scheme II

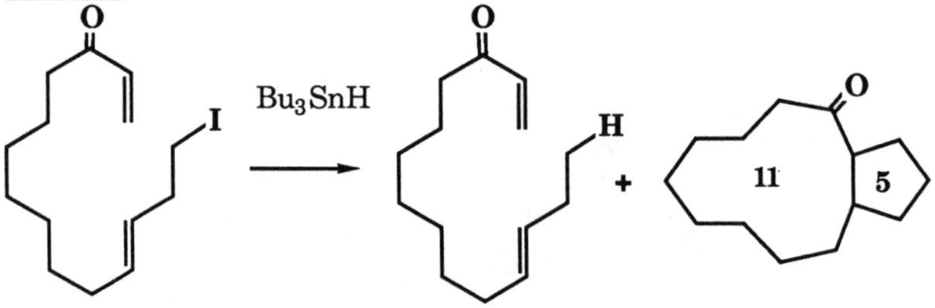

REFERENCES

1   Hart DJ (1984) Science 223:883
2   (a) Stork G, Sher PM, (1986) J Am Chem Soc 108:303, (b) Curran DP, Kuo SC (1986) J Am Chem Soc 108:1106
3   Tsang R, Dickson Jr J K, Pak H, Walton R, Fraser-Reid B (1987) J Am Chem Soc 109:3484
4   Beckwith ALJ, Moad G (1974) J Chem Soc Chem Comm 472
5   Griller D, Ingold KU (1980) Accs Chem Res 13:317
6   Beckwith ALJ, Schiesser CH (1985) Tetrahedron 41:3925
7   Porter NA, Magnin DR, Wright BT (1986) J Am Chem Soc 108:2787
8   Porter NA, Chang VHT (1987) J Am Chem Soc 109:4976
9   Porter NA, Chang VHT, Magnin DR, Wright BT (1988) J Am Chem Soc 110 in press
10   Porter NA, Lacher B, to be published

STEREOCHEMICAL CONTROL IN HEX-5-ENYL RADICAL CYCLIZATIONS

T. V. RAJANBABU

Central Research & Development Department
E. I. du Pont de Nemours and Co.
Experimental Station, Bldg. 328
Wilmington, Delaware 19898 U. S. A

The Wittig reaction of the sugars readily provide hex-5-ene-1-ols which were converted to hex-5-enyl radicals via one of the variations of the Barton deoxygenation reaction (equation 1)[1].

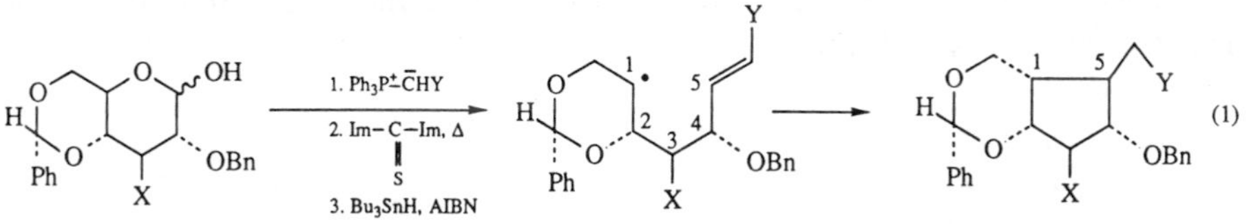

The stereochemistry of the newly formed carbon-carbon bond, i. e., 1,5-stereochemistry, is controlled primarily by the configuration of the C4 center of the radical. Unprecedented and exclusive 1,5-trans selectivity is realized in the gluco(1) series, whereas the manno(3) and galacto(4) systems lead to almost exclusive 1,5-cis stereochemistry. The C4-deoxy system gives a mixture of both 1,5-cis and trans products with the former predominating.

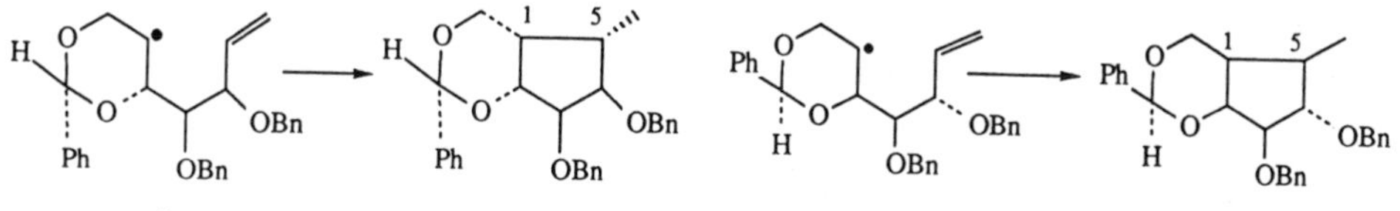

The stereochemical outcomes are rationalized by the cyclohexane-like transition state whose conformation[2], chair- or boat-like, is determined by both steric and stereoelctronic effects of substituent groups.

H. Fischer, H. Heimgartner (Eds.)
Organic Free Radicals
© Springer-Verlag Berlin Heidelberg 1988

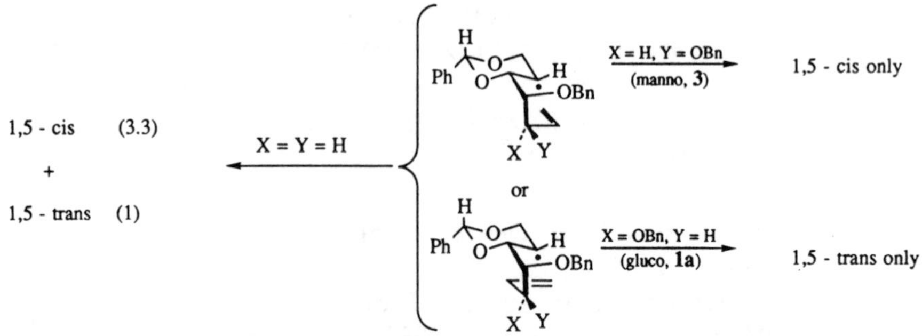

$$
\begin{cases}
\text{Ph} \diagdown \text{O} \diagdown \text{(manno, 3)} & X = H, Y = OBn \longrightarrow 1,5 \text{ - cis only} \\
\text{or} & \\
\text{Ph} \diagdown \text{O} \diagdown \text{(gluco, 1a)} & X = OBn, Y = H \longrightarrow 1,5 \text{ - trans only}
\end{cases}
$$

X = Y = H

1,5 - cis   (3.3)

+

1,5 - trans   (1)

For example, in these systems the favorable conformation of the C3-C6 segment(**5**) of the hexenyl chain which avoids 1,3-strain may be responsible for the seemingly high-energy boat-like transition state in the gluco-series[1]. A radical route to prostanoid intermediates including Corey lactone(**6**) makes use of the product **2 b**.

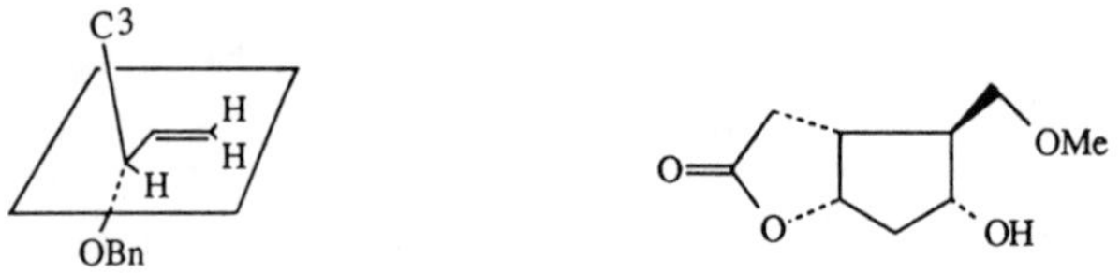

5

6

1. RajanBabu, T. V. *J. Am. Chem. Soc.* **1987,** *109*, 609.
2. (a) Beckwith, A. L. J.; Easton, C. J.; Lawrence, T.; Serelis, A. K. *Aust. J. Chem.* **1983**, *36*, 545. (b) Spellmeyer, D. C.; Houk, K. N. J. Org. Chem. **1987**, *52*, 959.

AN E.S.R. STUDY OF AMINIUM RADICAL
CATIONS WITH SILYL SUBSTITUTENTS

Christopher J. Rhodes

School of Chemistry
Thames Polytechnic
Wellington Street
Woolwich, London, SE18 6PF

ABSTRACT: The anisotropic ^{14}N couplings (2B) in silyl–aminium radical cations indicate appreciable delocation of spin–density onto the silicon substituents.

It is well known that silyl derivatives of main group elements have bonds between silicon and nitrogen, oxygen and fluorine which are short compared with the sums of the covalent radii[1]. The bond angles Si–O–Si and Si–N–Si in siloxanes[1] and silylamines[2] are also large compared with their organic counterparts, and, thus, in contrast with pyramidal organic amines, compounds such as $N(SiH_3)_3$ are actually planar at nitrogen[2].

This effect has most often been attributed to delocalisation of π-electron density into vacant silicon d-orbitals[3], and, more recently, to negative hyper-conjugation involving Si–C σ^* orbitals[3]. On the other hand, it has been proposed on the basis of the constancy of the α and β-proton couplings in a series of radicals, $R\dot{C}HX$ (where X is a group IV or V substitutent) that delocalisation of the unpaired electron onto the substitutent (X) is relatively unimportant[4].

With all this in mind, we decided to study a series of aminium radical cations with Me_3Si- substituents since couplings to the central (^{14}N) atom nuclei should be readily observed, and thus provide a measure of the spin–density at the radical centres, and hence the degree of delocalisation in these radicals.

Our results are shown in table 1, and demonstrate a clear reduction in the magnitude of the anisotropic ^{14}N coupling (2B), in accord with a delocalisation of 16% per Me_3Si-group. As in the case of alkyl radicals, the $\underline{a}(\alpha-H)$ couplings do not reflect this change, but it has been shown previously that the McConnel Q-value (for α-proton couplings) varies to some extent with the nature of the other substitutents at the radical centre[5], and so a compensating mechanism may operate, which partly offsets the reduction in $\underline{a}(\alpha-H)$ expected by spin delocalisation from the radical centre.

H. Fischer, H. Heimgartner (Eds.)
Organic Free Radicals
© Springer-Verlag Berlin Heidelberg 1988

The marked reduction which is observed in the methyl proton couplings in the series $MeN\overset{+\cdot}{H}_2$, $Me_2\overset{+\cdot}{N}H$, $Me_3\overset{+\cdot}{N}$, would seem to accord with a hyperconjugative model for β-proton couplings, in which increasing amounts of spin density are delocalised by the increasing number of methyl groups; however, the anisotropic ^{14}N couplings (2B) are hardly changed, and indicate that the spin-density at the central nitrogen atom is barely affected by increasing methyl substitution. This is a surprising contrast, and we are uncertain of the correct explanation for it. One possibility may be that the β-proton couplings in aminium cations arise predominantly from a spin-polarisation mechanism of the type proposed previously to account for part of the coupling to β-protons in alkyl radicals[7], although it was concluded that hyperconjugation was still operating to the extent of ca 50% In the aminium cations, hyperconjugation would need to make a much lower contribution than it does in alkyl radicals.

REFERENCES

(1) D.W.H. Rankin & H.E. Robertson, J. Chem. Soc.,Dalton Trans., 1983, 265
(2) K. Hedberg, J. Am. Chem. Soc., 1955, 77, 6491.
(3) A.E. Reed, P.V. Rague Schleyer, P. Vishnu Kamath & J. Chandrasekhar, J. Chem. Soc., Chem. Commun., 1988, 67, and references therein.
(4) A.R. Lyons, G.W. Neilson & M.C.R. Symons, J. Chem. Soc., Faraday Trans. 2, 1972, 807.
(5) H. Fischer, Z. Naturforsch., 1965, 20A, 428.
(6) V.N. Belevsky, L.T. Bugaenko & O. In Quan, J. Radioanal. Nucl. Chem., Letters, 1986, 107, 67.
(7) J.A. Brivati, K.D.J. Root, M.C.R. Symons & D.J.A. Tinling, J. Chem. Soc. (a), 1969, 1942.

Hyperfine Couplings (G)[a] in Aminium Radical Cations at 77 K

Radical	$^{14}N_{iso}$	2B	$A_{N\parallel}$	$A_{N\perp}$	$A_{N-H\parallel}$	$A_{N-H\perp}$	$A_{N-H_{iso}}$	A(Me)
[b] $MeN\overset{+\cdot}{H}_2$	20	29.0	49	5.5	–	–	22.5	46
[b] $Me_2\overset{+\cdot}{N}H$	19.3	29.7	49	4.3	36	19	24.7	34
[b] $Me_3\overset{+\cdot}{N}$	18.3	28.7	47	$0\overset{+}{-}4$	–	–	–	28.2
[c] $Me_3SiN\overset{+\cdot}{M}e_2$	16.0	24.0	40	$0\overset{+}{-}4$	–	–	–	31
[c] $(Me_3Si)_2\overset{+\cdot}{N}Me$	15.0	22.0	37	$0\overset{+}{-}4$	–	–	–	35
[c] $(Me_3Si)_2\overset{+\cdot}{N}H$	14.7	21.3	36	$0\overset{+}{-}4$	26	23	24	–

a, $1G = 10^{-4}T$; b, from ref. 6; c, this work.

THE CATION-RADICAL OF A HINDERED QUINONE DIAZIDE

H. Iwamura, A. Izuoka, T. Kohzuma, K. Ishiguro, Y. Sawaki,
I. Mitulidis, and A. Rieker

Institute for Molecular Science, Okazaki, 444 Japan,
Faculty of Engineering, Nagoya University, Chikusa-ku,
Nagoya, Japan,
and Institute for Organic Chemistry, University of Tübingen,
7400 Tübingen, Fed. Rep. Germany

Abstract: The anodic oxidation of 4-diazo-2,6-di-tert-butylcyclohexa-2,5-dienone at $-88^{\circ}C$ produces a cation-radical that has been shown by ESR and AM1 MO calculations to have a π-radical structure similar to that of phenoxy radicals but quite in contrast to the σ/π-radical nature of simple diazoalkane cation radicals.

Oxidation of diazoalkanes, especially using organic radicals as oxidants, has been assumed to proceed by one-electron oxidation via the corresponding cation radicals [1-3], although the presence of those species in solution could never be established.

Recently, we have been able to show by ESR spectroscopy at low temperatures that the electrochemical oxidation of phenyldiazomethanes generates the corresponding cation radicals in a σ- or π-state depending on the substituents [4]. Here, we report on the unique π-cation-radical **1a** of 4-diazo-2,6-di-tert-butylcyclohexa-2,5-dienone (**2a**), in which positive charge and odd electron are separated far apart.

The quinone diazide **2a** was prepared from the corresponding quinone by a one-pot modification of the known [5] procedure involving alkaline cleavage of the quinone tosylhydrazone. Electrolysis of **2a** in CH_2Cl_2 containing 0.1M Bu_4NBF_4 at $-88^{\circ}C$ under nitrogen in an ESR cavity at a potential of $+ 1.2 - 1.4$ V vs Ag/AgCl gave rise to a nine-line ESR spectrum corresponding to **1a**, which could be simulated with parameters $a_{N1(1N)} = 0.18$ mT; $a_{N2(1N)} = 0.36$ mT; $a_{H3(2H)} = 0.21$ mT.

H. Fischer, H. Heimgartner (Eds.)
Organic Free Radicals
© Springer-Verlag Berlin Heidelberg 1988

a: R = tBu
b: R = H

In sharp contrast with either the σ- or π-cation radicals of phenyldiazomethanes, the observed g-value (2.0051 ± 0.0001) is decidedly
larger than that of the free electron suggesting a phenoxy radical
structure for **1a**. This is further supported by the following arguments:

1. The corresponding phenoxy radical **3**, in which the diazonio group
is replaced by a cyano group shows a_N = 0.13 mT, a_{H3} = 0.22 mT [6].

2. AM1 MO calculations on **1b** and **2b** (optimized by assuming planar C_{2v}
structures), reveal for **1b** that (i) the net charge is very much
concentrated in the N_2 moiety, (ii) the spin density is mainly
distributed over the oxygen atom and C-2,4,6 of the ring, (iii) the
N^α-N^β bond is almost triple, the N^α-(C4) bond almost single, the bond
alternation in the ring being much reduced.

In contrast to the π-cation-radical structures (⊕ and ⊙ reside on
an allylic π-orbital perpendicular to the molecular plane) or σ-
cation-radical structures (⊕ resides on the allylic π-orbital, ⊙
localizes on the bent nitrogen atom) of other diazoalkanes, **1a** has
the fully separated phenoxyl and diazonium cation structure.

References:

1 Denny DB, Newman NF (1967) J Am Chem Soc 89:4692
2 Bethell D, Handoo KL, Fairhurst SA, Sutcliffe LH (1979) J Chem Soc,
  Perkin II 707
3 Winter W, Moosmayer A, Rieker A (1982) Z Naturforsch B 37:1623
4 Ishiguro K, Sawaki Y, Izuoka A, Sugawara T, Iwamura H (1987) J Am
  Chem Soc 109:2530
5 Nikiforov GA, Plekhanova LG, De Jonge K, Ershov VV (1978) Izv Akad
  Nauk, Ser Khim 2752
6 Rieker A (1961) Thesis University of Tübingen

FREE RADICAL FORMATION BY ULTRASOUND IN AQUEOUS SOLUTIONS OF VOLATILE AND NON-VOLATILE SOLUTES. A SPIN-TRAPPING STUDY

C. Murali Krishna, T. Kondo and P. Riesz

Radiation Oncology Branch, NCI
National Institutes of Health
Bethesda, MD 20892 USA

The creation, growth and collapse of small gas bubbles in a liquid exposed to ultrasound is called acoustic cavitation and is an effective mechanism for transforming the low-energy density of a sound field into the high-energy density characteristic of the interior and surrounding of a collapsing gas bubble. During the violent collapse of transient cavities, very high temperatures and pressures are produced which result in the thermal dissociation of water vapor into H atoms and OH radicals as the primary products of the sonolysis of water. Their formation has been confirmed recently in both continuous(1) and pulsed ultrasound(2). Reactions involving these radicals occur within the gas bubbles, at the liquid-gas interface and in the surrounding liquid. Reactions in the interior of the gas bubble are typical of those found in combustion, while those radicals that escape into the the bulk of the solution will react with other solutes as one would expect from radiation chemistry. The spin trap 3,5-dibromo-4-nitrosobenzene sulfonate (DBNBS) was used as the spin trap for sonolysis studies since the sulfonate group ensures non-volatility.

In this work, the sonochemically generated radicals from several non-volatile solutes (nucleic acid and protein constituents) and volatile solutes (alcohols) were identified by spin trapping with DBNBS and compared with those generated by H_2O_2-UV photolysis and gamma-radiolysis.

Non-Volatile solutes: The sonochemically generated spin adducts from argon-saturated aqueous solutions containing various pyrimidine bases, nucleosides(3), alkylpyrimidines(4), amino acids and dipeptides(5) were characterized and compared to those generated by gamma radiolysis of N_2O-saturated solutions or H_2O_2-UV photolysis of argon-saturated solutions. The spin trapped radicals produced by sonolysis were explained in terms of H atom and OH radical mediated H-abstraction reactions and H atom and OH radical addition to the C5-C6 double bond of pyrimidines.

H. Fischer, H. Heimgartner (Eds.)
Organic Free Radicals
© Springer-Verlag Berlin Heidelberg 1988

Volatile solutes: In the study of the sonolysis of aqueous solutions of volatile solutes, argon-saturated aqueous solutions of methanol(6), ethanol, 1-propanol, 2-propanol and t-butanol were investigated by ESR and spin trapping over the complete range of solvent composition. Methyl radicals and H-abstraction radicals were observed, as well as the isotopically mixed CH_2D and CHD_2 radicals when ROD:D_2O mixtures were studied. The results show that thermal decomposition of alcohols to methyl radicals and multiple radical recombination reactions occur in the gas phase. In the case of methanol-water mixtures, the methyl radical yield rises sharply at very low concentrations of methanol and reaches a maximum at 5 M mathanol in water and decreases with further increase in methanol concentration. The vapor pressure of the solution has been considered to have a significant influence on the sonochemical yields. The sonochemical yield of the spin adducts from water mixtures of the other alcohols as a function of solvent composition exhibited a more complex behavior.

Competitive scavenging studies:

The effect of various H atom and OH radical scavengers on the spin adducts of H and OH radicals was studied by Makino et al(1). The scavenging ability of certain volatile scavengers(acetone, 2-methyl-2-nitrosopropane) was found to be two orders of magnitude higher than would be expected from their known rate constants in homogeneous solutions. Henglein and Kormann(7) studied the effects of various solutes (volatile and non-volatile) on the H_2O_2 yield produced by sonolysis. The scavenging ability of these solutes was found to be related to their hydrophobicity but not to their rate constants with OH radicals. The Effects of non-volatile scavengers on OH radical induced thymine radicals(8) as measured by spin trapping were found to parallel their rate constants with OH radicals. This study indicates that the radicals produced from non-volatile solutes by ultrasound are formed in the bulk of the solution as well as in the interfacial region.

References:

1 Makino K, Mossoba MM, Riesz P (1983) J. Phys. Chem. 87,1369

2 Christman CL, Carmichael AJ, Mossoba MM, Riesz P (1987) Ultrasonics 25,31

3 Kondo T, Murali Krishna C, Riesz P (1988) Int. J. Radiat. Biol. 53,331

4 Kondo T, Murali Krishna C, Riesz P (1988) Radiat. Res (in press)

5 Murali Krishna C, Kondo T, Riesz P (1988) Radiat. Phys. Chem. 32,121

6 Murali Krishna C, Lion Y, Kondo T, Riesz P (1987) J. Phys. Chem. 91,5847

7 Henglein A, Kormann C (1985) Int. J. Radiat. Biol. 48,251

8 Kondo T, Murali Krishna C, Riesz P (1988) Int. J. Radiat. Biol (in press)

POLARITY REVERSAL CATALYSIS OF HYDROGEN
ATOM ABSTRACTION REACTIONS IN SOLUTION

Vikram Paul and Brian P Roberts*

Christopher Ingold Laboratories, University College
London, 20 Gordon Street, London WC1H 0AJ, U.K.

The importance of polar factors in influencing radical reactivity has, of course, been recognised for very many years.   In this poster, we describe how polar effects can be exploited to control the chemo- and regio-selectivities of hydrogen atom transfer processes in a *catalytic* manner.

Consider the abstraction of electron deficient hydrogen (eqn.1) and electron rich hydrogen (eqn.2) by an electrophilic radical $El^{\cdot}$ when both reactions are similarly exothermic.   In general, it is found that the activation energy for reaction (2) is significantly smaller than that for (1), because of the existence of a stabilising

$$El^{\cdot} \quad + \quad \overset{\delta+\;\delta-}{H-X} \quad \xrightarrow{\text{slow}} \quad El-H \quad + \quad X^{\cdot} \tag{1}$$

$$El^{\cdot} \quad + \quad \overset{\delta-\;\delta+}{H-Y} \quad \xrightarrow{\text{fast}} \quad El-H \quad + \quad Y^{\cdot} \tag{2}$$

$$[El^{\cdot} \; H-Y]^{\ddagger} \quad \longleftrightarrow \quad [El-H \; Y^{\cdot}]^{\ddagger} \quad \longleftrightarrow \quad [El^{\bar{\cdot}} \; H^{\cdot} \; Y^{+}]^{\ddagger}$$
$$\underline{(1a)} \qquad\qquad\qquad \underline{(1b)} \qquad\qquad\qquad \underline{(1c)}$$

charge-transfer interaction in the transition state for the former reaction, as represented by the inclusion of structure ($\underline{1c}$) in a valence bond description.   Charge transfer will not be important in the transition state for reaction (1) because of the relatively low stabilities of El^{+} and X^{+}.   For example, the activation energy for reaction (3) is appreciably smaller than for reaction (4), despite the fact that both abstractions are exothermic by *ca.* 50 kJ mol^{-1}.

$$Bu^{t}O^{\cdot} \quad + \quad CH_{3}CN \quad \xrightarrow{\text{slow}} \quad Bu^{t}OH \quad + \quad {}^{\cdot}CH_{2}CN \tag{3}$$

$$Bu^{t}O^{\cdot} \quad + \quad CH_{3}OBu^{t} \quad \xrightarrow{\text{fast}} \quad Bu^{t}OH \quad + \quad {}^{\cdot}CH_{2}OBu^{t} \tag{4}$$

H. Fischer, H. Heimgartner (Eds.)
Organic Free Radicals
© Springer-Verlag Berlin Heidelberg 1988

However, the overall reaction (3) can be *catalysed* if the single step is replaced by the pair of consecutive reactions (5) and (6). Now the *donor catalyst* H-Nuc reacts rapidly (because of favourable polar effects) to furnish the *nucleophilic* radical Nuc$^\bullet$, which in turn rapidly abstracts the electron deficient hydrogen from

$$Bu^tO^\bullet \ + \ \overset{\delta- \ \delta+}{H-Nuc} \ \xrightarrow{\text{fast}} \ Bu^tOH \ + \ Nuc^\bullet \tag{5}$$

$$Nuc^\bullet \ + \ CH_3CN \ \xrightarrow{\text{fast}} \ H-Nuc \ + \ {}^\bullet CH_2CN \tag{6}$$

acetonitrile to regenerate H-Nuc.

We have already demonstrated[1] the practical viability of this principle of *polarity reversal catalysis* (PRC) for controlling homolytic chemoselectivity by using amine-alkylboranes (*e.g.* $Me_3N{\rightarrow}BH_2Thx$; Thx = $-CMe_2CMe_2H$) as a donor catalyst for reaction (3). In the poster we describe further applications of PRC as a tool for control of chemo- and regio-selectivity. For example, t-butoxyl radicals react with bis(2-cyanoethyl) ether to give exclusively the α-oxyalkyl radical (2), while in the presence of a small amount of $Me_3N{\rightarrow}BH_2Thx$ overall abstraction of electron deficient hydrogen to yield the α-cyanoalkyl radical (3) takes place.

$$Bu^tO^\bullet \ + \ (NCCH_2CH_2)_2O \ \begin{array}{l} \xrightarrow{\text{without PRC}} \ NCCH_2\overset{\bullet}{C}HOCH_2CH_2CN \ (\underline{2}) \\ \\ \xrightarrow{+ \ Me_3N{\rightarrow}BH_2Thx} \ NC\overset{\bullet}{C}HCH_2OCH_2CH_2CN \ (\underline{3}) \end{array}$$

At low temperature t-butoxyl radicals undergo preferential *addition* to cyclopenta-1,3-diene, while in the presence of $Me_3N{\rightarrow}BH_2Thx$ as polarity reversal catalyst, hydrogen atom abstraction to give the cyclopentadienyl radical occurs exclusively.

We note that the principle of PRC should be equally applicable to sluggish abstraction of electron rich hydrogen by nucleophilic radicals, when an *acceptor catalyst* H-El will be required.

1 Paul V, Roberts BP (1987) J Chem Soc Chem Commun 1322

KINETICS AND MECHANISM OF THE OXIDATION
REACTION OF THIOETHERS BY CERIUM(IV) AMMONIUM NITRATE

E. Baciocchi[a], D. Intini[b], and C. Rol[b]

[a]Dipartimento di Chimica, Università "La Sapienza", 00185 Roma, Italy
[b]Dipartimento di Chimica, Università di Perugia, 06100 Perugia, Italy

One electron oxidation of organic sulfides is certainly an important
process particularly for the role that it may have in several bio-
chemical transformations. In spite of this, detailed studies on
chemically induced one electron transfer reactions of thioethers are
very scarce (1).

The reaction of cerium(IV) ammonium nitrate (CAN) with di-n-butyl
sulfide, diphenyl sulfide, thioanisole and a series of aryl benzyl
sulfides in AcOH leads to the corresponding sulfoxides almost
quantitatively. When the reaction is carried out at 25°C using a
CAN:substrate molar ratio of 2:1, two moles of CAN react with one
mole of sulfide to give one mole of sulfoxide.

Working in the experimental conditions where this stoichiometry is
observed, the reaction is first order in both reagents and the reaction
rate is slightly decreased when cerium(III) nitrate is added to the
reaction mixture. Moreover the reaction shows a high sensitivity to
substrate structure, a ρ (σ^+) value of -3.3 being calculated from
reactivity data for a series of aryl benzyl sulfides ($XC_6H_4SCH_2C_6H_5$ with
$X=p-CH_3,H,p-Cl,m-Cl$). Good correlations are observed plotting the second
order rate constants (k_2) against both anodic peak potentials (E_p) of
the same substrates in CH_3CN and charge transfer transition energies
($h\nu_{CT}$) of the corresponding TCNE complexes in CCl_4, with the following
slopes: $\Delta log\ k_2/\Delta E_p=-0.55$ and $\Delta log\ k_2/\Delta h\nu_{CT}=-0.54$. Noteworthy is the
higher reactivity of thioanisole respect to di-n-butyl sulfide, in line
with ionization potential data ($E_p,^{h\nu}_{CT}$) and in contrast with

H. Fischer, H. Heimgartner (Eds.)
Organic Free Radicals
© Springer-Verlag Berlin Heidelberg 1988

relative reactivity observed in reactions operating via an electrophilic attack of oxidant at sulfur.

These observations suggest the reversible formation of a radical cation which then might be attacked by a nitrate ion or acetic acid to form sulfoxide.

Work is in progress to determine the actual oxygen source for sulfoxide formation.

1 Riley DP, Smith MR, Correa PE (1988) J Am Chem Soc 110:177; Gilmore JR, Mellor JM (1971) Tetrahedron Lett 3977; Watanabe Y, Iyanagi T, Oae S (1980) Tetrahedron Lett 3685; Srinivasan C, Chellamani A, Rajagopal S (1985) J Org Chem 50:1201

NUCLEOPHILIC SUBSTITUTION IN α-FUNCTIONALIZED TERTIARY HALOALKANES BY RESONANCE STABILIZED CARBANIONS

Francisco Ros[*] and José de la Rosa

Instituto de Química Médica, CSIC,
Juan de la Cierva 3, 28006 Madrid, Spain

Relatively few cases of C-C bond forming $S_{RN}1$ substitutions at a saturated carbon atom of substrates lacking a nitro group are known. Examples of this kind are the C-alkylation of nitroalkane anions by α-chloroisobutyrophenone (1) or α-bromoisobutyronitrile (2). In relation to these reactions we have examined those of compounds **1** with nitroalkane, α-carboxyethyl-α-cyano, or α-cyanobenzylic carbanions.

$$R\!-\!\underset{\underset{CH_3}{|}}{\overset{\overset{CH_3}{|}}{C}}\!-\!X \qquad\qquad EtO_2C\!-\!\underset{\underset{CH_3}{|}}{\overset{\overset{C\equiv N}{|}}{C}}\!-\!X$$

1

X = Cl; R = PhCO   X = Cl, Br

X = Br; R = PhCO, CO_2Et, C≡N

From α-bromoisobutyrophenone, or ethyl α-bromoisobutyrate, and nitroalkane anions low yields of $S_{RN}1$ C-alkylation products (**2**) are obtained; products resulting from ionic reactions are also formed (**3**, **4**). For these systems the electron transfer propagating step of the $S_{RN}1$ sequence may be dissociative.

$$R\!-\!\underset{\underset{CH_3}{|}}{\overset{\overset{CH_3}{|}}{C}}\!-\!A \qquad\qquad EtO_2C\!-\!\underset{\underset{CH_3}{|}}{\overset{\overset{C\equiv N}{|}}{C}}\!-\!\underset{\underset{R''}{|}}{\overset{\overset{R'}{|}}{C}}\!-\!NO_2$$

2

R = PhCO, CO_2Et, C≡N

A = $O_2NCR'R''$, $EtO_2CC(C\!\equiv\!N)R'$,

$\quad$ PhC(C≡N)R', $Ph_2CC\!\equiv\!N$

H. Fischer, H. Heimgartner (Eds.)
Organic Free Radicals
© Springer-Verlag Berlin Heidelberg 1988

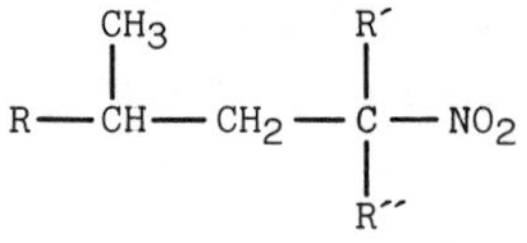

$$\underset{\textbf{3}}{R-\underset{\underset{}{}}{\overset{CH_3}{\overset{|}{CH}}}-CH_2-\underset{\underset{R''}{|}}{\overset{\overset{R'}{|}}{C}}-NO_2} \qquad\qquad \underset{\textbf{4}}{R-\underset{\underset{CH_3}{|}}{\overset{\overset{CH_3}{|}}{C}}-OH}$$

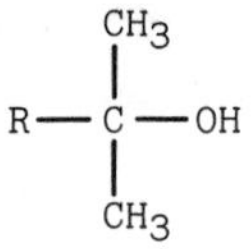

R = PhCO, CO$_2$Et   R = PhCO, CO$_2$Et

$$\underset{\textbf{5}}{[O_2NC(CH_3)_2-]_2} \qquad\qquad \underset{\textbf{6}}{[EtO_2CC(C\equiv N)(CH_3)-]_2}$$

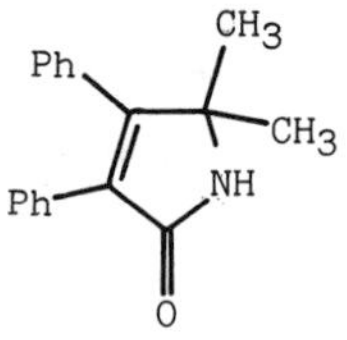

7

Extensive dimerization (products **5** and/or **6**) occurs during the alkylations of nitroalkane anions by ethyl α-cyano-α-halopropionates. In this case the cross coupling products appear to be formed by S$_N$2 displacement at the halogen atom of the substrates followed by a S$_{RN}$1 reaction between the resulting <u>gem</u>-halonitroalkane and α-cyano carbanion.

Cyclic products such as **7** are formed in the reactions of α-cyano carbanions with α-chloroisobutyrophenone, while with the α-bromo compounds the substitution products (**2**) are readily yielded by reactions which do not involve radical chains.

References

1 Russell GA, Ros F (1985) J Am Chem Soc 107:2506
2 Ros F, de la Rosa J (1988) J Org Chem 53, in press

PHOTOEXCITED ELECTRON TRANSFER AND SOLVENT-DEPENDENT PATH WAYS OF SPIN POLARIZATION AND RADICAL FORMATION IN SOLUTIONS WITH ANTHRAQUINONE

H.-K. Roth and D. Leopold

Department of Natural Sciences, Leipzig University of
Technology, PSF 66, DDR-7030 Leipzig, GDR

A.A. Obynochny, Yu. P. Zentalovich, E.G. Bagryanskaya,
Y.A. Grishin and R. Z. Sagdeev

Institute of Chemical Kinetics and Combustion
SU-630090 Novosibirsk 90, USSR

ABSTRACT

The path ways of photolytically induced radical formation and of first steps of radical reacions have been studied by various NMR and ESR methods including time-domain measurements in the time scale of one microsecond and the detection of radical pairs by stimulated nuclear polarization, SNP. The generation of chemically induced polarization of electron spins and nuclear spins was used as the main surce of information about the fast laser-induced processes in solutions of anthraquinone, AQ, and triethylamine, TEA, in various polar and nonpolar solvents.

ESR AND ESE STUDIES

The main information of ESR time-profile measurements is that the radical precursors are triplet state molecules. The signals are strongly emissive polarized due to the radical creation from a tripet state whose higher T_{+1} levels are much more populated than those in the thermal equilibrium. Spectra recorded in the time-window of (1-3)μs after the laser flashes of an excimer laser operating at 308 nm belong to the anion radical AQ^{-} of anthraquinone formed in solutions with TEA as follows /1/:

$$^{1}AQ \xrightarrow{h\nu} \, ^{1}AQ^{*} \xrightarrow{\text{isc}} \, ^{3}AQ^{*} \qquad (1)$$

$$^{3}AQ^{*} + \, ^{1}NEt_{3} \longrightarrow \, ^{2}AQ^{-} + \, ^{\dagger}NEt_{3} \qquad (2)$$

By ESR the cation radical of TEA could not be detected. In solution of AQ in isopropanol and benzene without TEA the neutral semiquinone radical AQH was detected in ESE experiments.

$$^{3}AQ^{*} + LH \longrightarrow A\dot{Q}H + L^{\bullet} \qquad (3)$$

LH - solvent molecule, $L^{\bullet}$ - solvent radical

AQ^{-} and $A\dot{Q}H$ have different relaxation times and recombination rates /1,2/.

SNP STUDIES

The stimulated nuclear polarization technique allows the measurement of ESR data of radical pairs by NMR polarization measurement /3,4,5/. The photolysis takes place in an ESR magnetic field of 0,05 T.

H. Fischer, H. Heimgartner (Eds.)
Organic Free Radicals
© Springer-Verlag Berlin Heidelberg 1988

The RF-field of ESR stimulated the S-To mixing thus a different amount of polarized reaction products is to be observed in NMR measurements.

In acetonitrile and methanol the spectra of the reaction products as well those of solvent molecules taking part in the exchange of protons (or deuterons) show well resolved SNP-ESR spectra of the cation radical $^+NEt_3$. The hyperfine constant amounts to about 2 mT in accordance with the ODESR results. The measurement in various solvents and the SNP observations of different molecule groups yield a different sign of polarization behaviour, from which we get information on the path ways of reaction.

CIDNP STUDIES AND REACTION PATH WAYS

In the case of the polar solvents acetonitrile and methanol the ion radical pairs have relatively long life-times and can be polarized by the RPM. The CIDNP spectra show emission of the CH_2-protons of TEA and no polarization of the methyl protons. In the strong polar solvents the reaction takes place only via the ion path way according to equation (2). In addition to the polarization on products we observe enhanced absorption of OH-protons on methanol and of water which is solved in CD_3OD. The solvent molecules LD can participate in the polarization according to following equations:

$$AQ^{\mp} + LD + \updownarrow NEt_3 \longrightarrow A\dot{Q}D + L^- + \updownarrow NEt_3 \qquad (4)$$

$$L^- + \updownarrow NEt_3 \longrightarrow LH + CH_3-\dot{C}H-NEt_2 \qquad (5)$$

In the case of the nonpolar solvents C_6D_6 and C_6D_{12} the ion radical pairs have only a very short life-time. For that reason the polarization generation takes place via neutral radicals. [$A\dot{Q}H$, $CH_3-\dot{C}H-NEt_2$] are the predominant radical pairs and on TEA the polarization can be observed mainly on the CH_3 group. In addition, polarization appears on diethylvinylamine and on other products of escaping radical pairs.

The dielectric constant of isopropanol is situated between the strong polar and nonpolar solvents. The pure radical path way as well as the ion-radical path way of polarization generation occur.

From the CIDNP time-domain measurements we know that the cage processes takes place (in all solvents used) in times shorter than 1 μs.

In addition to the polarity of the solvents also their viscosity and the concentration of AQ and TEA determine the rate of polarization generation in various solvents and on different molecule groups.

References

1   Roth HK, Leopold D (1988) Makromol Chem/Macromol Symp in press
2   Beckert D, Schneider G (1987) Chem Phys 116:421
3   Bagryanskaya EG, Grishin YuA, Sagdeev RZ, Leshina TV,
    Polyakov NE, Molin YuA (1985) Chem Phys letters 117:220
4   Bagryanskaya EG, Grishin YuA, Sagdeev RZ, Molin YuN
    Chem Phys letters 128:417
5   Mikhailov SA, Salikhov KM and Plato M (1987) Chem Phys 117:197

PHOTOCHEMICALLY GENERATED RADICALS
IN INDUSTRIAL APPLICATIONS

W. Rutsch

CIBA-GEIGY AG

Additives Research Department

CH-1701 Freiburg, Switzerland

Modern technologies are characterized by high quality products, efficient and economical production processes as well as a high regard for safety and ecology. UV-curing, which makes use of photochemically generated free radicals to crosslink coatings and printing inks, is a typical representative of this technology. In an introduction to the lecture the most important industrial applications will be presented.

The compounds used to produce these initiating radicals photochemically are called photoinitiators. They have found a high level of commercial acceptance since their industrial introduction in the early 1970's. We are in the situation today of having available resins and photoinitiators which meet to a large extent the technical requirements of the coatings industry. Such systems show the required stability and reactivity, and also yield coatings that have good adhesion, gloss, abrasion resistance, flexibility and are non-yellowing.

Aryl-alkyl-ketones, having various substituents on the α-carbon atom and on the benzene ring, are the most important class of photoinitiators (see Figure 1). They have desirable absorption characteristics in the range of 300–400 nm, and undergo efficient photochemical reactions. The region of 300–400 nm is of special significance because clear coatings are transparent to light at these wavelengths, and this light can be produced efficiently with medium pressure mercury lamps which have wide industrial acceptance.

$$\text{I:}\quad R_1 = -H,\ R_2 = R_3 = -OCH_3,\ R_4 = -C_6H_5 \tag{1}$$

$$\text{II:}\quad R_1 = -H,\ R_2 = R_3 = -(CH_2)_5,\ R_4 = -OH \tag{2}$$

$$\text{III:}\quad R_1 = -SCH_3,\ R_2 = R_3 = -CH_3,\ R_4 = -N\!\!\bigcirc\!\!O \tag{3}$$

Figure 1: Alkyl-aryl-ketones: important classes of photoinitiators

Application results in clear coatings and pigmented systems will be presented to demonstrate the dramatic changes in performance that can be obtained by varying the R_1–R_4 substituents in the photoinitiator structure (4).

H. Fischer, H. Heimgartner (Eds.)
Organic Free Radicals
© Springer-Verlag Berlin Heidelberg 1988

The elucidation of the involved photochemical reaction mechanisms has challenged many research groups. Their findings will be used to interpret the application results (5):

- ^{1}H-NMR CIDNP and trapping experiments show that the initiating radicals are predominantly produced by Norrish-type I cleavage (α-cleavage) from a first excited triplet state with predominant n$\rightarrow \pi^*$ character.

- Triplet energies and triplet lifetimes are discussed to explain reactivity differences in various monomers and reactive diluents (quenching effects).

- An increase in UV-absorption caused, for example, by a sulfur substituent in the para position of the benzoyl chromophore allows efficient utilization of the available light if thin films (3-5 μm) or pigmented systems have to be cured.

- Sensitizers can increase the efficiency of selected photoinitiators and extend their sensitivity towards longer wavelengths. Evidence for triplet sensitized α-cleavage is presented for the combination of thioxanthones with 2-methyl-1-[4-(methylthio)phenyl]-2-morpholino-propan-1-one.

1 Kirchmayr R, Berner G, Rist G (1980) Farbe & Lack 86:224
2 Kirchmayr R, Berner G, Hüsler R, Rist G (1982) Farbe & Lack 88:910
3 Rutsch W, Berner G, Kirchmayr R, Hüsler R, Rist G, Bühler N (1986) Org Coat 8:175
4 Desobry V, Dietliker K, Hüsler R, Rutsch W, Löliger H (1987) Eurocoat 87 Nice (France) XVII Congress of the AFTPV-Peintures Conf Proc p 96
5 Hageman HJ (1985) Progr Org Coat 13:123 and references cited therein

LOW-FIELD CIDNP MANIFESTATION OF
THE MINOR RECOMBINATION CHANNEL
VIA TRIPLET STATE OF RADICAL PAIRS

E.C.Korolenko, K.M.Salikhov and N.V.Shokhirev

Institute of Chemical Kinetics & Combustion, USSR Academy of
Sciences, Siberian Branch, 630090, Novosibirsk, USSR.

Net low-field CIDNP effects arising in the recombination
product of a one-nuclear radical pair (RP) with spin I= 1/2 have
been calculated numerically. It is supposed, that besides a major
RP recombination channel via singlet (S) state, a minor triplet (T)
recombination channel exists. It is shown, that the total (S+T)
CIDNP sign can be determined by the minor T-recombination channel.
This effect takes place if the RP in-cage lifetime and the total
residence time in the reaction zone are sufficiently long.

The motion of the radicals is described according to the
continuous diffusional model. The stochastic Liouville equation for
the RP spin density matrix is solved by the finite difference
method. Uncharged radicals and radical-ion pairs are considered. The
influence of the exchange interaction is studied. The following
parameters have been varied: the solvent dielectrical permittivity
constant (ε), the radicals' mutual diffusion coefficient (D), the
exchange integral magnitude and the recombination rates of singlet
(W_S) and triplet (W_T) RPs.

Figs. 1-3 show the results of the calculations. The analysis of
these leads to the conclusion, that for long-lived RPs, the observed
CIDNP is contributed predominantly by the minor T-recombination
channel. The simplified RP model calculations also confirm this
conclusion. The results obtained show, that the low-field CIDNP
experimenting can reveal the minor RP T-recombination channel.

References

1 Mikhailov SA, Salikhov KM, Plato M (1987) Chem Phys 117:197
2 Korolenko EC, Shokhirev NV, Leshina TV (1986) The calculation of
  CIDNP in radical-ion pairs by the finite difference method,
  Pre-print of the Institute of Chemical Kinetics & Combustion,
  Novosibirsk

H. Fischer, H. Heimgartner (Eds.)
Organic Free Radicals
© Springer-Verlag Berlin Heidelberg 1988

Fig. 1. CIDNP dependence on dielectrical permittivity constant
(ε) for a radical-ion pair. At
strong Coulomb attraction, the
total CIDNP sign is determined by
the minor T-recombination channel.

    Curve 1: CIDNP in S-product,

    Curve 2: CIDNP in T-product,

    Curve 3: total (S+T) CIDNP.

Parameters used: T-precursor,

H_o= 2 mT, A= 2 mT, W_S= 10^9 nm/s,
W_T= 10^7 nm/s, J=0, D= 10^{-5} cm^2/s.

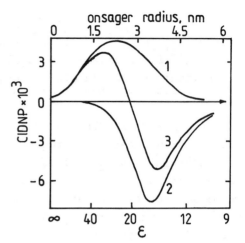

Fig. 2. CIDNP dependence on the
diffusion coefficient (D) for
uncharged radicals. At high vis-
cosities, the total CIDNP sign is
determined by the minor T-recombination channel.

    Curve 1: CIDNP in S-product,

    Curve 2: CIDNP in T-product,

    Curve 3: total (S+T) CIDNP.

Parameters used: T-precursor,

H_o= 2 mT, A= 2 mT, W_S= 10^9 nm/s,
W_T= 10^7 nm/s, J=0.

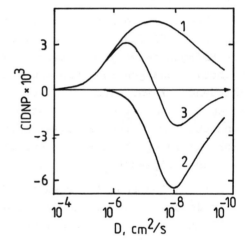

Fig. 3. Influence of the minor
T-channel recombination rate (W_T)
on the total CIDNP dependence on
ε. With decreasing W_T, the
minor T-channel reveals itself at
smaller ε values (i.e. stronger
Coulomb attraction).

    Curve 1: W_T= 10^8 nm/s,

    Curve 2: W_T= 10^7 nm/s,

    Curve 3: W_T= 10^6 nm/s.

Parameters used: T-precursor,

H_o=0.5 mT, A= 2 mT, W_S= 10^9 nm/s,
J=0, D= 10^{-5} cm^2/s.

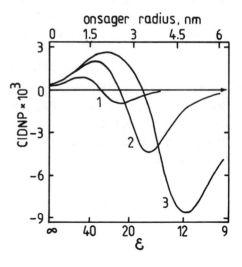

CHLORINE-DIOXIDE AS AN ELECTRON TRANSFER OXIDANT OF OLEFINS

S. Sarel, Ch. Rav-Acha and E. Choshen

The Hebrew University of Jerusalem, Israel

Kinetic and product studies of chlorine-dioxide oxidation of olefins, ranging in their ionization potentials from 9.64 ± 0.3 to 8.0 eV have shown that at the rate-determing stage, the reaction involved the abstraction of an electron from the olefin to yield the anion ClO_2^{-} with generation of the corresponding cation radical $\overset{+}{\underset{}{C}} = \overset{\cdot}{C}$. The products emerging from the oxidation in H_2O were suggested to arise from an hydrolysis reaction of a postulated cyclic-intermediate.

We wish to demonstrate that the above sequence: (i) ionization, (ii) pericyclic reaction, (iii) electron acceptance could be applied successfully to the ClO_2-oxidation of stilbene in CCl_4 as delineated below:

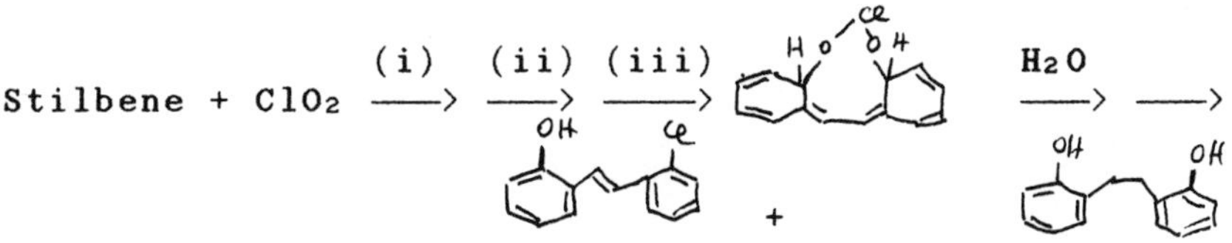

Stilbene + $ClO_2 \xrightarrow{(i)} \xrightarrow{(ii)} \xrightarrow{(iii)} \quad \cdots \quad \xrightarrow{H_2O} \longrightarrow$

62%   +   17%

(1) C. Rav-Acha, E. Choshen, and S. Sarel, Helv. Chem. Acta., 1986, 69, 1728.

(2) B.D. Lindgren and T. Nilsson, Acta Chem. Scand., 1974, B28, 847.

H. Fischer, H. Heimgartner (Eds.)
Organic Free Radicals
© Springer-Verlag Berlin Heidelberg 1988

REDOX-ACTIVE IRON-CHELATORS BASED ON PYRIDOXAL

S. Avramovici-Grisaru and S. Sarel

The Hebrew University School of Pharmacy, P.O.B. 12065,

Jerusalem, Israel

Electron spin resonance (ESR) and cyclicvoltametric (CV) studies indicate that pyridoxal-based iron chelators of general structure (I) are capable, through metal sequestration (I—>II), of transfering single electrons from ferrous ions (II—III) towards the electron-deficient center to yield a free-radical of presumed structure (III). In presence of appropriate electron acceptors [Acc], (III) converts to (IV) with concommitant formation of the corresponding anion radical [Acc]$^{\doteq}$. In absence of suitable aₙ electron-acceptor, the intermediacy of a nitrogen-centrered free radical is involved to rationalize the observed generation of double-decomposition products[1]. Data in substantiation of this view will be presented.

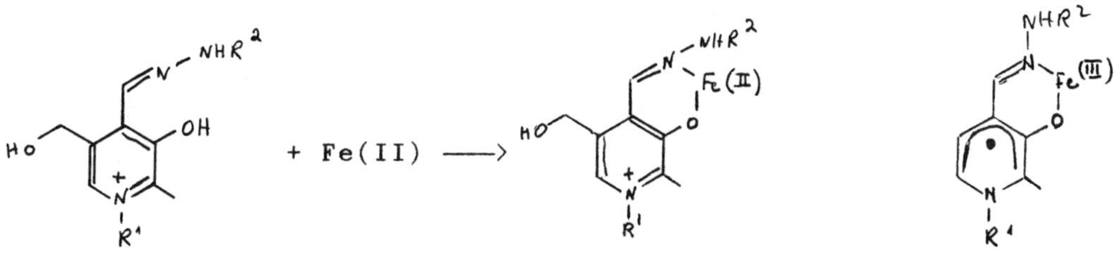

$$(I) \qquad + \; Fe(II) \longrightarrow \qquad (II) \qquad\qquad (III)$$

H. Fischer, H. Heimgartner (Eds.)
Organic Free Radicals
© Springer-Verlag Berlin Heidelberg 1988

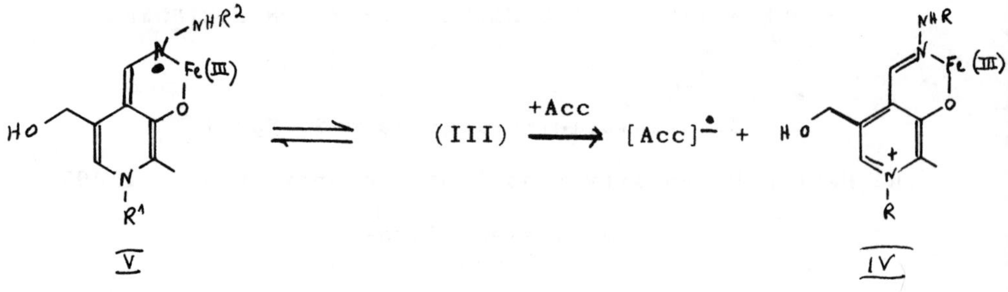

$$(III) \xrightarrow{+Acc} [Acc]^{\bullet -} +$$

1) S. Avramovici-Grisaru and S. Sarel, <u>Chem. Comm.</u> 1986.

TWO PHOTON PROCESSES IN
FREE RADICAL CHEMISTRY

J.C. Scaiano

Division of Chemistry, National Research Council
Ottawa, Canada K1A 0R6

Two-laser, two-color techniques have been used to study reactions involving sequential absorption of two photons (1,2). These processes can be divided in two groups: (A) Two-photon processes leading to ground state free radicals, and (B) Generation of free radicals in monophotonic processes, followed by laser excitation of the radical into an electronic excited state. The developments in these areas are summarized below.

(B) TWO PHOTON SOURCES OF FREE RADICALS.

A rather common situation in photochemistry is the case where, following excitation, a molecule undergoes rapid intersystem crossing to a triplet state which has less energy than that required to cleave the weakest bond in the molecule. As a result, the molecule can be photochemically inert towards intramolecular decomposition (Figure 1).

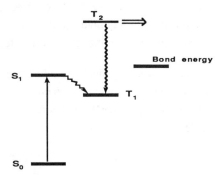

Figure1: Jablonsky diagram showing the energy levels of various electronic states and the weakest bond energy in a molecule with suitable properties as a laser specific photoinitiator.

However, if the triplet state (T_1) in Fig. 1 is reexcited by a second photon, sufficient energy can become available for cleavage to occur. These molecules are effectively laser specific photoinitiators. Molecules such as benzil (3), bromoarenes and some monoketones (4) show this type of behavior. This clearly implies that the upper excited states (e.g. T_2) are sufficiently long lived that chemical reactions can compete with internal conversion (5).

H. Fischer, H. Heimgartner (Eds.)
Organic Free Radicals
© Springer-Verlag Berlin Heidelberg 1988

(B) EXCITED STATES OF FREE RADICALS.

Excited free radicals have been generated in two-laser experiments. In this case the first laser is used as a *synthesis laser*, followed by a second, delayed laser, of a wavelength suitable to excite the radicals with no or minimum photodecomposition of the free radical precursor (6). Several benzylic radicals, including diphenylmethyl (6), phenanthrylmethyl (7), naphthylmethyl (7,8) and biphenylmethyl (7) have been generated using this technique. These excited radicals generally fluoresce in the 500-600 nm region (6-9) and in some cases have remarkably long lifetimes in solution at room temperature; for example for 2-phenanthrylmethyl and diphenylmethyl the lifetimes are 79 and 250 ns, respectively (6,7). Excited benzylic radicals are excellent electron donors and moderate to poor electron acceptors. They interact readily with oxygen, but these processes do not lead to peroxyl radicals; we have suggested that the process may lead to singlet oxygen ($^1O_2$) formation via energy transfer from the excited radical (6). The excited radicals are remarkably good halogen abstractors, a reaction that in some cases proceeds even with Freons. Hydrogen abstraction is not a preferred pathway for excited radicals; for example $Ph_2CH\bullet^*$ reacts rather sluggishly with Bu_3SnH, the excited species being rather long lived (170 ns) even when the tin hydride is used as solvent (6).

Diaryl ketyl radicals have also been examined in detail. Their lifetimes are frequently in the 2-10 ns range at room temperature and are strongly dependent on substitution at the O-H position (10). The decay of $Ph_2COH\bullet^*$ involves efficient cleavage of the O-H bond (11), a process that in toluene occurs with a quantum yield of 0.27. The lifetimes of excited ketyl radicals are substantially longer in conformationally restricted molecules.

The spectroscopy of excited biradicals is very similar to that of free radicals with the same radical centers (12) ; however, the excited state lifetimes can be much shorter in the case of biradicals (E.g. 2.5 ns for $\bullet Ph_2C(CH_2)_3CPh_2\bullet^*$) and the chemistry quite different from that observed for monoradicals

REFERENCES.

(1) Scaiano JC, Johnston LJ (1986) Pure Appl.Chem. 58:1273.
(2) Scaiano JC, Johnston LJ, McGimpsey WG, Weir D (1988) Acc.Chem.Res. 21:22.
(3) McGimpsey WG, Scaiano JC (1987) J.Am.Chem.Soc. 109:2179.
(4) Johnston LJ, Scaiano JC (1987) J.Am.Chem.Soc. 109:5487.
(5) McGimpsey WG, Scaiano JC (1988) J.Am.Chem.Soc. 110:2299
(6) Scaiano JC, Tanner M, Weir D (1985) J.Am.Chem.Soc. 107:4396.
(7) Weir D, Johnston LJ, Scaiano JC (1988) J.Phys.Chem. 92:1742.
(8) Johnston LJ, Scaiano JC (1985) J.Am.Chem.Soc. 107:6368.
(9) Weir D, Scaiano JC (1986) Chem.Phys.Lett. 128:156
(10) Johnston LJ, Lougnot DJ, Scaiano JC (1986) Chem.Phys.Lett. 129:205.
(11) Johnston LJ, Lougnot DJ, Wintgens V, Scaiano JC (1988) J.Am.Chem.Soc. 110:518.
(12) Johnston LJ, Scaiano JC (1986) J.Am.Chem.Soc. 108:2349.

TETRAHYDROFURANS, PYRROLIDINES AND CYCLOPENTANES BY INTRAMOLECULAR CYCLIZATION OF KOLBE-RADICALS

H.J. Schäfer, L. Feldhues and G. Dralle

Organisch–Chemisches Institut der Universität,
Correns–Str. 40, 4400 Münster, FRG

Abstract: Tetrahydrofurans, pyrrolidines and cyclopentanes are synthesized by intramolecular addition of 5–hexenyl radicals generated by Kolbe–electrolysis from carboxylates.

Five membered carbocycles[1] and heterocycles[2] can be effectively prepared by a 5–*exo*–trig–cyclization via radicals mostly generated from bromides. We have shown, that radicals produced from carboxylic acids by Kolbe–electrolysis can also be used for this cyclization (eq. 1). The reaction proceeds by anodic decarboxylation to a 1–hexenyl radical that undergoes a 5–*exo*–trig–cyclization to a primary alkyl radical that couples with another radical generated by coelectrolysis from a second acid.

(1)

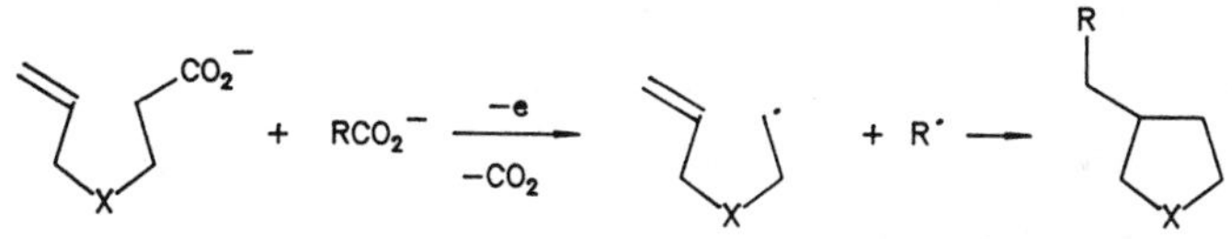

X: O, NCOR , C(CH₃)₂

β–Allyloxypropionates, prepared by Michael–addition of allylalcohols to acrylonitrile, are coelectrolyzed with carboxylic acids to afford substituted tetrahydrofurans in up to 70 % yield (eq. 2)[3].

(2)

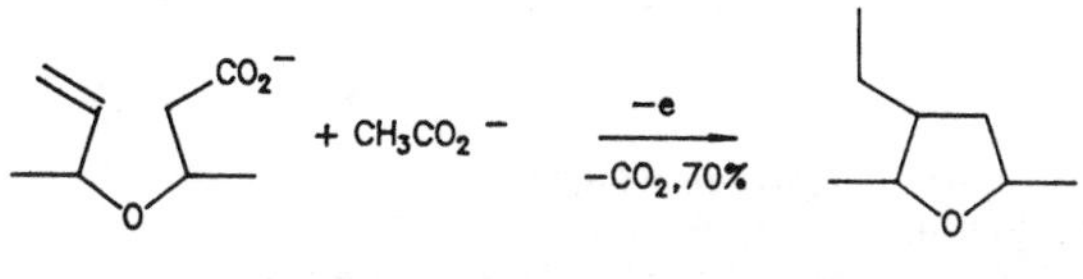

Inspired by the prostaglandine synthesis of *Stork*[4], which uses a radical cyclization, we applied the Kolbe–electrolysis to this approach. Target molecule was the acetal 1, which allows further conversion to prostaglandines of type $PGF_{2\alpha}$ (eq. 3)[5].

H. Fischer, H. Heimgartner (Eds.)
Organic Free Radicals
© Springer-Verlag Berlin Heidelberg 1988

(3)

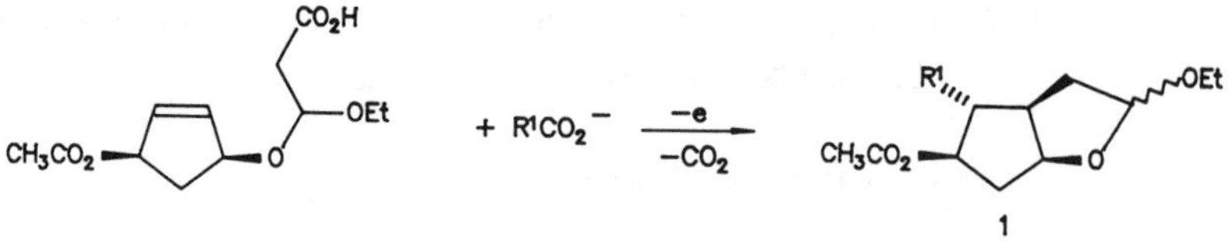

3-Alkyl-substituted pyrrolidines are obtained in 45 to 67 % yield by Kolbe-electrolysis of β-allylaminoalkanoates and mixed coupling of the cyclized radical with the radical of a coacid (eq. 4)[6].

(4)

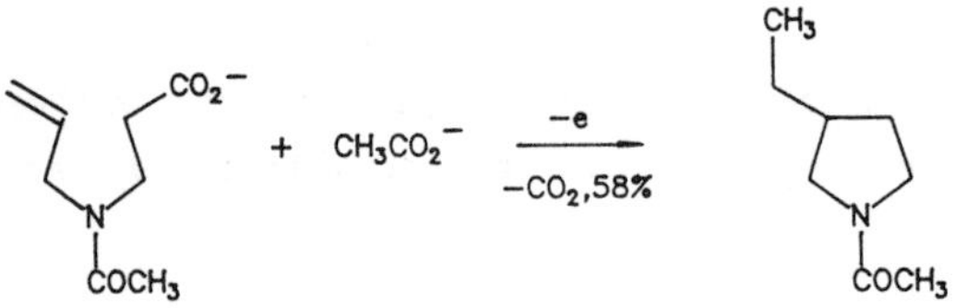

The coelecrolysis of 4,4-dimethyl-6-heptanoate with methylsuccinate, methyladipate or hexanoate leads to cyclic and acyclic coupling products in a ratio of about 1:1 (eq. 5).

(5)

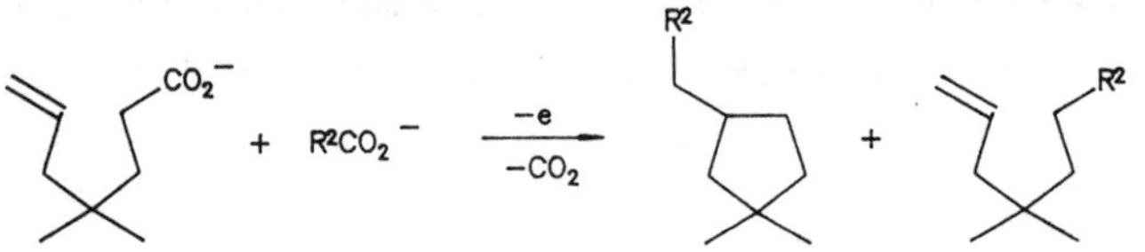

R^2: $(CH_2)_2CO_2Me$, $(CH_2)_4CO_2Me$, $(CH_2)_5CH_3$

The intramolecular Kolbe-addition has the major advantage compared with the radical chain addition starting from the bromide, that *two* C-C bonds are being formed, whilst in the latter only *one* C-C bond and *one* C-H bond are joined in most cases. Furthermore the second carbon substituent can be varied simply and in a wide range by the choice of the coacid in the mixed Kolbe-electrolysis.

References:
1 Curran DP, Kuo SC (1986) J Am Chem Soc 108:1106;
  Beckwith ALJ, Robert DH, Schiesser CH, Wallner A (1985)
  Tetrahedron Lett 3349; Stork G, Baine NH (1985) Tetrahedron
  Lett 5927.
2 Hart DJ (1984) Science 883; Ueno Y, Chino K (1982) J Am Chem Soc
  104:5564; Stork G, Kahn M (1985) J Am Chem Soc 107:500.
3 Huhtasaari M, Schäfer H J, Becking L (1984) Angew Chem IE Engl 23:980.
4 Stork G, Sher PM, Chen HL (1986) J Am Chem Soc 108:6384.
5,6 Becking L, Schäfer HJ (1988) Tetrahedron Lett in press

VITAMIN B_{12}, CATALYST OF RADICAL REACTIONS
IN ORGANIC SYNTHESIS

R. Scheffold, S. Busato, E. Eichenberger, R. Härter, H. Su, O. Tinembart, L. Walder
Ch. Weymuth, Z.-D. Zhang

Institute of Organic Chemistry, University of Bern
Freiestrasse 3, CH-3012 Bern

Vitamin B_{12} and related metal complexes act as mediators in the electron transfer
from the electron source (chemical reducing agent or cathode in electrochemical
reductions) to an electrophilic substrate R-X (R-X = alkyl-, vinyl- or acyl- derivative) [1]. On B_{12}-catalyzed reduction R-X is transformed to the radical R· or the
carbanion R⁻ (depending on the structure of R and the reaction conditions) either
directly or *via* an organometallic intermediate. The Co,C-bond of the organocobalamin
intermediate is cleaved as a consequence of redox-reactions [2], thermic or photochemical activation [3].

The vitamin B_{12}-catalyzed electrolysis of R-X at ca. -0.9 V *vs.*SCE under simultaneous
irradiation of visible light [4] is the mildest way to form radicals R· by *Umpolung*
of R-X. The thus created radicals may undergo follow-up reactions as e.g.:
intra- or intermolecular addition to olefins [5], elimination of β-leaving groups [6],
reduction followed by protonation [7], disproportionation to olefin and hydrocarbon
[8] or rearrangements [9].

To the synthetically most useful reactions belongs the the C,C-bond formation by
inter- or intramolecular addition to the (activated) double or triple bond of alkenes
or alkines as well as the (concerted) formation of more than one C,C-bond by subsequent inter- or intramolecular additions (*tandem*-reaction).

The application of the Organic Photo Electro Catalysis (OPEC-reaction) with B_{12} as
catalyst is illustrated by the synthesis of some natural products.

B_{12}-catalyzed intermolecular addition [1, 3, 4]

As an example the synthesis of all stereoisomers of brevicomin (an insect pheromon)
starting from cheap enantiomerically pure compounds like tartaric acid, mannitol,
vitamin C and ribonolactone is described [10]. The strategy for the construction of
bicyclic ketales is outlined in the scheme.

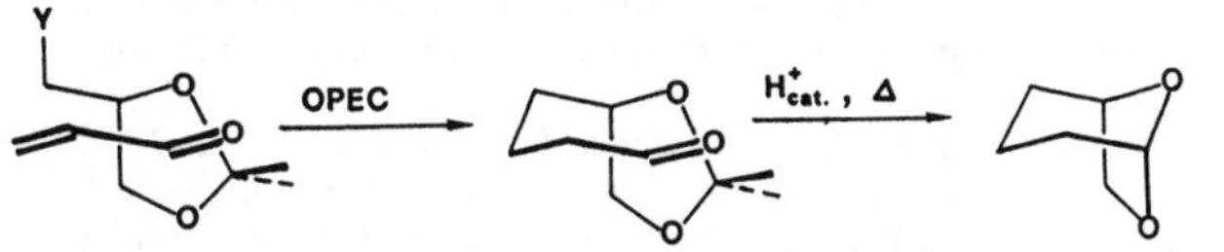

OPEC-reaction:

Electrolysis at -0.9 V *vs.*SCE
in 0.5 N LiClO$_4$/DMF in presence of ca. 1 mol % hydroxo-
cobalamine and irradiation
of visible light.

H. Fischer, H. Heimgartner (Eds.)
Organic Free Radicals
© Springer-Verlag Berlin Heidelberg 1988

B_{12}-catalyzed intramolecular addition [1, 2, 5, for radical cyclization cf.also 11].

The B_{12}-catalyzed cyclization with C,C-bond formation has first been described in 1980 [5]. It might be applied to systems in which the attack of the potential radical to a C,C-multiple bond occurs by a 5-, 6- or 7-*exo*- and a 6- or 7-*endo*-mode. As an example the synthesis of a bicyclic lactol (an intermediate in prostaglandin synthesis [12]) is described (scheme) [13].

B_{12}-catalyzed *tandem*-reaction [1a]

The consecutive formation of more than one C,C-bond in one operation may occur, if a set of structural elements like R-X and more than one C,C-multiple bond are properly arranged for inter- and intramolecular addition. The three possibilities for the formation of two C,C-bonds are: intra- followed by inter-, intra- followed by intra- and inter- followed by intramolecular addition. B_{12}-catalyzed *tandem*-reactions have been applied in the synthesis of brasilenol [14](for an independent synthesis cf.[15]) and the key-intermediate of $PGF_{2\alpha}$ [13](for related syntheses cf. [16])(scheme).

Asymmetric catalysis by vitamin B_{12} [17]

Recent results of B_{12}-catalyzed *in vitro* isomerization of epoxides are presented.

References:
[1] Reviews: a) R. Scheffold, Nachr.Chem.Tech.Lab. 36, 261 (1988); b) R. Scheffold
    S. Abrecht, R. Orlinski, H.R. Ruf, P. Stamouli, O. Tinembart, L. Walder, Ch.
    Weymuth,Pure & Appl.Chem. 59,363(1987);c) R.Scheffold, Chimia 39, 203 (1985).
[2] R. Scheffold, G. Rytz, L. Walder in "Modern Synthetic Methods" Ed. R. Scheffold,
    Vol. 3: Salle, Frankfurt/ Sauerländer, Aarau/ Wiley, London p. 355-440, 1983.
[3] L. Walder, R. Orlinski, Organometallics 6, 1606 (1987).
[4] R. Scheffold, R. Orlinski, J.Am.Chem.Soc. 102, 7200 (1983).
[5] R. Scheffold, M. Dike, S. Dike, T. Herold, L. Walder, J.Am.Chem.Soc.102,3642(1980).
[6] R. Scheffold, E. Amble, Angew.Chem. 92, 643 (1980); Int.Ed. 19, 629 (1980).
[7] L. Walder, G. Rytz, K. Meier, R. Scheffold, Helv.Chim.Acta 61, 3013 (1978).
[8] H. Su, R. Scheffold, unpublished; to bee publ. in Helv.Chim.Acta (1988).
[9] A. Fischli, T.S. Wan, Helv.Chim.Acta 67, 684, 1461, 1883 (1984).
[10] S. Husi, Ch. Weymuth, R. Scheffold, unpubl.; to be publ. in Helv.Chim.Acta (1988).
[11] a) B. Giese "Radicals in Organic Synthesis: Formation of Carbon-Carbon Bonds,
    Org. Chem. Series, Vol. 5, Pergamon Press, Oxford 1986; G. Stork, Bull.Chem.Soc.
    Jpn. 61, 149 (1988).
[12] J.J. Partridge, N.K. Chadha, M.R. Uskokovic, J.Am.Chem.Soc. 95, 7171 (1973).
[13] S. Busato, O. Tinembart, R. Scheffold, unpubl.; to be publ. in Helv.Chim.Acta.
[14] R. Härter, Ch. Weymuth, R. Scheffold, unpubl.; to be publ. in Helv.Chim.Acta.
[15] A.E. Greene, A.A. Serra, E.J. Barreiro, P.R.R. Costa, J.Org.Chem. 51, 4250 (1987).
[16] G.E. Keck, D.A. Burnett, J.Org.Chem. 52, 2959 (1987); G. Stork, P.M. Sher,
    H.-L. Chen, J.Am.Chem.Soc. 108, 6384 (1986).
[17] H. Su, L. Walder, Z.-D. Zhang, R. Scheffold, Helv.Chim.Acta, in print (1988).

SOLUTION CHEMISTRY OF KETO
AND ENOL CATION RADICALS

U. Baumann and M. Schmittel

Institut für Organische Chemie
Albertstr. 21, D-7800 Freiburg

The gas phase chemistry of keto and enol cation radicals has been extensively investigated by several groups using mass spectrometric techniques[1]. Numerous MO calculations and thermochemical studies[2,3] demonstrate that simple enol cation radicals are significantly more stable than their corresponding keto forms. A recent study[4] on 2,2-dimesitylethenol and 2,2-dimesitylethanal has provided a detailed picture of the gas phase ion chemistry, including keto enol tautomerization and specific rearrangements.

Using stable simple enols and the corresponding ketones we have studied the solution chemistry of the cation radicals. Cyclic voltammetry in various solvents at different temperatures revealed irreversible voltammograms of the ECE type indicative of fast subsequent reactions. To study the intrinsic reactivity of the cation radicals in more detail we have oxidized the ketones and enols in $CHClF_2$ and acetonitrile at different temperatures applying one electron oxidants like O_2AsF_6 and various aminium salts.

Mechanisms will be proposed to account for the formed products. Comparison with the gas phase ion chemistry discloses interesting differences in reactivity patterns.

References:

1 Biali SE, Depke G, Rappoport Z, Schwarz H (1984) J Am Chem Soc 106:496 and references quoted therein
2 Heinrich N, Koch W, Frenking G, Schwarz H (1986) J Am Chem Soc 108:593 and cited references
3 Holmes JL, Lossing FP (1980) J Am Chem Soc 102:1591
4 Rabin I, Biali SE, Rappoport Z, Lifshitz C (1986) Int J Mass Spectrom Ion Processes 70:301.

H. Fischer, H. Heimgartner (Eds.)
Organic Free Radicals
© Springer-Verlag Berlin Heidelberg 1988

HYDROXYL RADICAL AND LASER LIGHT-INDUCED STRAND BREAK FORMATION OF POLYNUCLEOTIDES AND DNA

D. Schulte-Frohlinde

Max-Planck-Institut für Strahlenchemie, Stiftstraße 34-36,
D-4330 Mülheim a.d. Ruhr

Abstract: Damage to DNA is the main cause of γ-irradiation-induced cell deactivation. The DNA damage in cells occurs either via OH radicals (indirect effect, contribution 50-80 %) or via radical cations and excess electron produced by ionization of the DNA (direct effect, contribution 20-50 %). The two effects can be studied independently in aqueous solution, the indirect effect by investigating the reaction of OH radicals with DNA and model-compounds in aqueous solution, the direct effect by high-intensity laser-induced ionization of DNA in aqueous solution. We studied the reaction of OH radicals with nucleobases, nucleosides and poly-nucleotides from the pyrimidine series (uracil, thymine, cytosine). The OH radicals add preliminarily to the C=C double bond or abstract a H atom from a methyl group.[1] The reaction of pyrimidine polyribo-nucleotides with OH radicals lead mainly to base adduct radical.[2] However, these base radicals abstract H atoms intramolecularly from the sugar.[3] H abstraction at the 2' or 4' position of the sugar leads to single-strand break (ssb) formation by heterolytic cleavage of the sugar phosphate bond in poly(U),[4] poly(C)[5] as well as in DNA.[6] Increase of proton concentration increases the rate of ssb formation by three orders of magnitude,[4] whereas addition of Mg^{+++} decreases the rate.
With the corresponding purine base (adenine, guanine) the reaction with OH radicals lead to fast elimination of OH^- and the intermediate formation of radical cations which convert to neutral radicals by subsequent release of a proton.[7] Detailed reaction schemes with rate constants of various steps have been obtained.[7,8] With polydeoxy-riboadenylic acid (poly(dA)) base radicals are intermediates in the course of the OH radical-induced ssb formation.[9] Since in poly(A) two pathways to ssb formation are available (the C-2' and the C-4' pathway) and in poly(dA) only one (the C-4' pathway), but the yield of ssb formation with poly(dA) is much greater than that with poly(A),

H. Fischer, H. Heimgartner (Eds.)
Organic Free Radicals
© Springer-Verlag Berlin Heidelberg 1988

the conformation and the structure of the polymers have a large influence on the yield.[9] The results are in line with a more flexible structure of poly(dA) in comparison to poly(A) which latter shows stronger base stacking and a different sugar puckering.

The direct effect was studied by ionizing nucleobases with laser pulses of high intensity in aqueous solution. Up to now a detailed free radical pathway leading to ssb induced by a laser pulse has been identified only for polyuridylic acid.[10] Results concerning the formation of strand breaks in DNA induced by laser excitation will be presented.[11] Aspects of the direct effect may also be studied by reaction with SO_4^- radical anions, since certain aromatic compounds generate radical cations by reaction with SO_4^-. The influence of SO_4^- on the biological activity of plasmid DNA was determined.[12] With uridine (uracilyl-riboside) as substrate the reaction of SO_4^- leads to sugar radicals whereas with OH radicals only base adduct radicals are observed.[13,14] Surprisingly with SO_4^- and deoxyuridine no sugar radicals are observed although the only difference between uridine and deoxyuridine is the presence of an OH substituent in position 2' of the sugar in uridine which is absent in deoxyuridine.[14] This result is explained by different sugar puckering in the two compounds (3'-endo in uridine, 2'-endo in deoxyuridine).

References:

1 Fujita S, Steenken S (1981) J Am Chem Soc 103:2540
  Hazra DK, Steenken S (1983) J Am Chem Soc 105:4380
  Deeble DJ, von Sonntag C (1986) Int J Radiat Biol 49:927
2 Deeble DJ, Schultz D, von Sonntag C (1986) Int J Radiat Biol 49:915
3 Lemaire DGE, Bothe E, Schulte-Frohlinde D (1984) Int J Radiat Biol 45:351
4 Schulte-Frohlinde D, Bothe E (1982) Z Naturforsch 37c:1191
5 Müller M (1983) Dissertation Ruhr-Universität Bochum
6 Dizdaroglu M, von Sonntag C, Schulte-Frohlinde D (1975) J Am Chem Soc 97:2277
7 Vieira AJSC, Steenken S (1987) J Am Chem Soc 109:7441
8 Vieira AJSC, Steenken S (1987) J Phys Chem 91:4138
9 Adinarayana M, Bothe E, Schulte-Frohlinde D (1988) Int J Radiat Biol (submitted)
10 Schulte-Frohlinde D, Opitz J, Görner H, Bothe E (1985) Int J Radiat Biol 48:397
11 Opitz J, Schulte-Frohlinde D (1987) J Photochem 39:145
12 Aboul-Enein A, Schulte-Frohlinde D (1988) Photochem Photobiol 47:
13 Behrens G, Herak JN, Hildenbrand K, Schulte-Frohlinde D (1988) J Chem Soc Perkin Trans 2 (submitted)
14 Schulte-Frohlinde D, Hildenbrand K (May 1988) Nato Advanced Study Institute Bardonlino Italy

ROLE OF NITRATE RADICAL IN THE NITROOXYLATION OF ALKENES BY CERIUM(IV) AMMONIUM NITRATE

E. Baciocchi[a], T. Del Giacco[b], S.M. Murgia[b], and G.V. Sebastiani[b]

[a]Dipartimento di Chimica, Università "La Sapienza", 00185 Roma, Italy
[b]Dipartimento di Chimica, Università di Perugia, 06100 Perugia, Italy

The photochemical reaction of cerium(IV)ammonium nitrate (CAN) in acetonitrile (irradiation with a high pressure mercury lamp by pyrex) with cyclohexene, 1-octene and a series of styrene derivatives leads to the formation of 1,2-dinitrate adducts in good to high yields (60-100%) under very mild conditions.

$$RCH=CH_2 \xrightarrow[\text{MeCN,r.t.}]{\text{CAN},h\nu} RCH(ONO_2)CH_2ONO_2$$

As previously reported (1,2,3), the nitrate radical formed in the photolysis of CAN, is the actual oxidizing species (eq. 1).

$$Ce^{IV}ONO_2 \xrightarrow{h\nu} Ce^{III} + NO_3^{\cdot} \qquad (1)$$

The reaction of $NO_3^{\cdot}$ with alkenes, which has been studied kinetically by the laser flash photolysis technique (ruby laser, λ_{exc}=347 nm), is a very fast process with k values between 5×10^8 and $9 \times 10^9 M^{-1}s^{-1}$.

The kinetic data for styrene derivatives fit in with the Rehm-Weller equation for electron transfer processes, thus suggesting an electron transfer mechanism. A cation radical intermediate is formed which then gives a ß-nitrate carbon radical (eqs. 2 and 3). The latter intermediate undergoes a ligand transfer with another CAN molecule to give the dinitrate adduct (eq. 4).

H. Fischer, H. Heimgartner (Eds.)
Organic Free Radicals
© Springer-Verlag Berlin Heidelberg 1988

$$ArCH=CH_2 + NO_3^{\cdot} \longrightarrow Ar\overset{+}{C}H\text{-}\overset{\cdot}{C}H_2 + NO_3^{-} \qquad (2)$$

$$Ar\overset{+}{C}H\text{-}\overset{\cdot}{C}H_2 + NO_3^{-} \longrightarrow Ar\overset{\cdot}{C}HCH_2ONO_2 \qquad (3)$$

$$Ar\overset{\cdot}{C}HCH_2ONO_2 + Ce^{IV}ONO_2 \longrightarrow ArCH(ONO_2)CH_2ONO_2 + Ce^{III} \quad (4)$$

This suggestion is also supported by the observation that trans-β-methylstyrene is more reactive than α-methylstyrene, in line with the easier oxidizability of the former substrate and in contrast with what is observed in free radical additions.

The kinetic data for 1-octene and cyclohexene do not fit the Rehm-Weller plot and the mechanistic attribution to the reaction of these substrates with $NO_3^{\cdot}$ is still uncertain.

1 Baciocchi E, Del Giacco T, Rol C, Sebastiani GV (1985) Tetrahedron Lett 26:541
2 Baciocchi E, Del Giacco T, Rol C, Sebastiani GV (1985) Tetrahedron Lett 26:3353
3 Baciocchi E, Del Giacco T, Murgia SM, Sebastiani GV (1987) J Chem Soc Chem Commun 1246

RING INVERSION IN DIOXENE
RADICAL CATIONS

C.J. Shields and A.G. Davies

University College London, 20 Gordon Street,
London WC1H OAJ, U.K.

No e.s.r. spectra of the radical cations of dioxenes or of other vinyl ethers have been reported in the literature.

We have prepared the dioxenes (1) and (2) by a novel synthetic route. The corresponding radical cations were generated by oxidation with several reagents, and the e.s.r. spectra were observed to be temperature dependent.

Dioxene rings are twisted into a half chair conformation. At high temperature the half chair rapidly interconverts ($A \rightleftharpoons B$), time-averaging the e.s.r. coupling from the pseudo-axial and pseudo-equatorial methylene hydrogens into a quintet. Lowering the temperature reduces the rate of interconversion until ultimately the triplet couplings are resolved.

We have studied by e.s.r. the interconversion rates of the dioxene radical cations as a function of temperature. From this we have established that the activation barrier (E_a) to ring inversion in the radical cations is 8.2 and 6.9 kcal mol^{-1} for (1^{+}) in $Hg(CF_3CO_2)_2 / CF_3CO_2H / CH_2Cl_2$ and $HFSO_3 / SO_2$ respectively, and 5.0 kcal mol^{-1} for (2^{+}) in $Hg(CF_3CO_2)_2 / CF_3CO_2H / CH_2Cl_2$. These values will be compared with the activation barriers to ring inversion in the neutral dioxenes as determined by variable temperature n.m.r.[1].

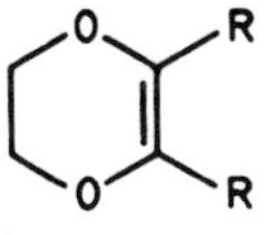

(1) R=CH$_3$

(2) R=H   (A)   (B)

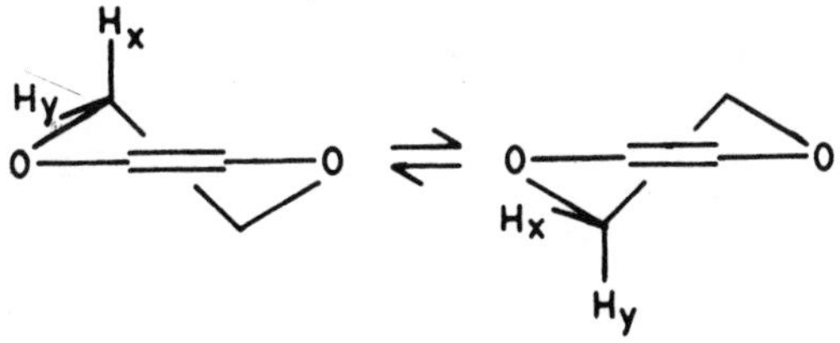

(1) R.H. Larkin and R.C. Lord, J. Am. Chem. Soc., 1973, 95, 5129

H. Fischer, H. Heimgartner (Eds.)
Organic Free Radicals
© Springer-Verlag Berlin Heidelberg 1988

ANTIOXIDANT SYSTEMS
IN BIOLOGICAL MATERIAL

H. Sies

Institut für Physiologische Chemie I
Universität Düsseldorf
Moorenstrasse 5, D-4000 Düsseldorf 1

The shift in the balance in prooxidant/antioxidant activities in cells has been termed oxidative stress (see 1). Multiple lines of antioxidant defense exist in cells and organs as well as body fluids. Antioxidants include activities in prevention, in interception and repair. All major classes of biological molecules are susceptible to oxidative damage, so that DNA, proteins, lipids and carbohydrates may be attacked. A recent review deals with these types of damage and also with the antioxidant defense (2).

NON-ENZYMATIC ANTIOXIDANTS

Some vitamins play an important role in this regard. Vitamin E, vitamin C and beta-carotene (provitamin A) carry the predominant activity (3). The formation of the chromanoxyl radical from alpha-tocopherols occurs with a rate constant of about 10^6 $M^{-1}s^{-1}$ (4). The regeneration of alpha-tocopherol can occur with vitamin C, forming the ascorbyl radical, or with glutathione, forming the glutathionyl radical (5,6). The antioxidant capacity of beta-carotene, also known as provitamin A, is based on its efficiency as a quencher of singlet molecular oxygen (7,8). Most of the effects of beta-carotene and other carotenoids can be explained by this reaction.
The protection of biological membranes against lipid peroxidation by ascorbate or glutathione depends on the presence of tocopherol in the membrane (9).

ENZYMATIC ANTIOXIDANTS

This field is characterized by three major activities, superoxide dismutase, glutathione peroxidases and catalase. These react in complementary ways, both as to specificity of reactions catalyzed and to subcellular localization (10). In addition, a number of socalled ancillary enzymes were described, including conjugation enzymes like

H. Fischer, H. Heimgartner (Eds.)
Organic Free Radicals
© Springer-Verlag Berlin Heidelberg 1988

quinone reductase, glutathione transferases, glucuronyl transferases etc, as well as transport systems that export products of free radical metabolism (2).

The control of the levels of enzymes has been studied in different systems. In microorganisms it was shown that the oxidative damage to DNA leading to spontaneous mutagenesis is under the control of a regulon, oxyR, and deletions of this regulon lead to higher frequencies of spontaneous mutagenesis (11). Further, in eukaryotic cells, DNA hypomethylation was shown to increase some of the activities of the ancillary antioxidant enzymes (12).

SYNTHETIC ANTIOXIDANTS

In addition to plant products like flavonoids, there is a number of chemically synthesized free radical scavenging antioxidants such as butylated hydroxytoloene and butylated hydroxyanisole. We have been recently interested in a selenoorganic compound, ebselen, which catalyses the reaction of GSH with hydroperoxides, e.e. the glutathione peroxidase reaction (13). The compound ebselen is also active in protecting against lipid peroxidation in intact cells, whereas it does not protect in glutathione-depleted cells; this led to the conclusion that ebselen protects against membrane damage by its GSH peroxidase reaction.

1 Sies H (ed) (1985) Oxidative stress. Academic Press, London
2 Sies H (1986) Angew Chem Int Ed 25:1058
3 Burton GW, Cheeseman KH, Doba T, Ingold KU, Slater TF (1983) In: Biology of vitamin E. Pitman, London
4 Burton GW, Ingold KU (1981) J Am Chem Soc 103:6472
5 Niki E (1987) Ann N Y Acad Sci 498:186
6 McCay PB (1985) Ann Rev Nutr 5:323
7 Foote CS, Denny RW (1968) J Am Chem Soc 90:6233
8 Krinsky NI (1979) Pure & Appl Chem 51:649
9 Wefers H, Sies H (1988) Eur J Biochem in press
10 Chance B, Sies H, Boveris A (1979) Physiol Rev 59:527
11 Storz G, Christman MF, Sies H, Ames BN (1987) Proc Natl Acad Sci USA 84:8917
12 Wagner G, Pott U, Bruckschen M, Sies H (1988) Biochem J in press
13 Müller A, Cadenas E, Graf P, Sies H (1984) Biochem Pharmacol 33:3235

CYCLIZING BIFUNCTIONNALIZATION
OF DIENES BY SULFONYL HALIDES

I. De Riggi, J.-M. Surzur and M.P. Bertrand

Laboratoire de Chimie Organique B, associé au CNRS (n° 109)
Faculté des Sciences (St Jérôme) - 13397 Marseille Cedex 13 - France

Free radical sulfonyl halides addition to olefins is well documented. In the same way free radical addition to non conjugated dienes such as *cis, cis* -1,5 cyclooctadiene or diallylic compounds is probably one of the first known example of free radical cyclization giving yield to monomeric compounds (1) or to the cyclopolymerization process (2).

We describe here the first examples, to our knowledge, of free radical addition-cyclization of sulfonyl halides to some non conjugated olefins.

The same trends as in the 5-hexenyl radical cyclization are observed : exclusive five membered ring formation, preferential formation of the *cis* -compound. In fact from 1,5- *cis, cis* cyclooctadiene only one bifunctionnalized cyclized product is isolated presumably the one shown in scheme 1. But the formation of this compound is quite sensitive to the experimental conditions and to the nature of the X group : if the transfer rate is higher than the cyclization rate only the uncyclized product is obtained.

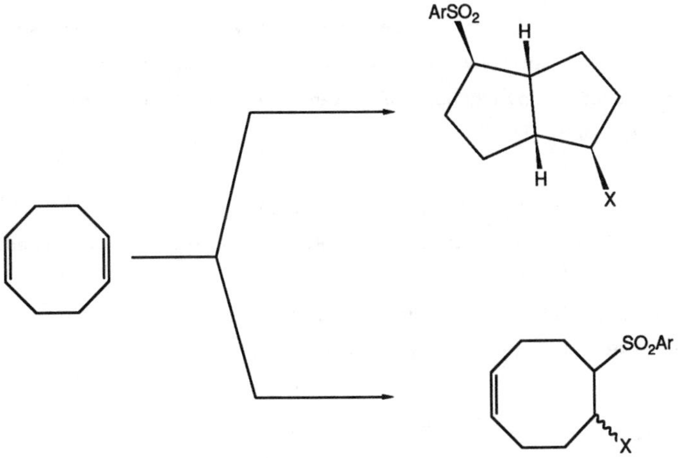

Scheme 1

H. Fischer, H. Heimgartner (Eds.)
Organic Free Radicals
© Springer-Verlag Berlin Heidelberg 1988

With diallylic compounds the stereoselectivity for the *cis* product is higher than the one observed with C-1 monosubstituted 5-hexenyl radicals generally 2.3:1 (3) beeing 3:1 (G = N–Ts) or 4:1 (G = O). When G = SiPh$_2$ the six membered ring is now selectively formed in accordance with the results observed from β-sila-5-hexen-1-yl radicals (4). Only the *cis* isomer is observed.

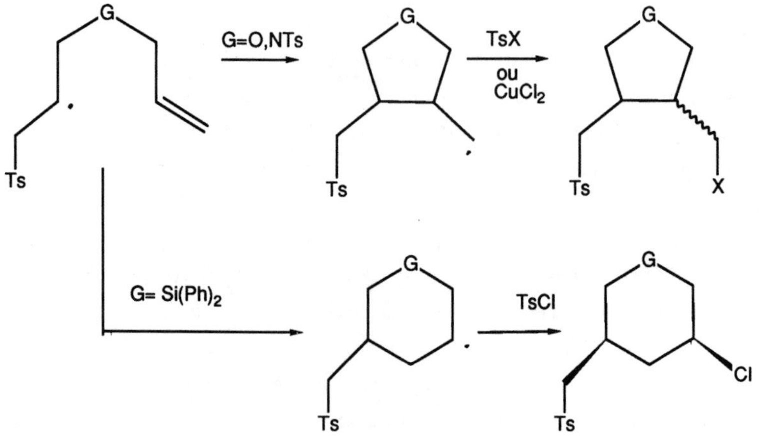

Scheme 2

Yields are of course very dependent on the experimental conditions but although they have not been optimized they can be as high as 74%.

Since the starting materials are easily prepared and because of the potentialities of the sulfonyl and halide groups, the products obtained can be considered as useful synthons to elaborate more complex carbocyclic and heterocyclic compounds.

(1) Butler G.B., Acc. Chem. Res. (1982), 15 : 370.
(2) Surzur J.-M. (1982) In : Abramovitch R.A. (ed) Reactive intermediates, vol 2, Plenum Press, New York.
(3) Beckwith A.L.J. (1981) Tetrahedron 37 : 3073.
(4) Wilt J.W. (1985) Tetrahedron 41 : 3979.

RADICAL STRUCTURE IN   - IRRADIATED SINGLE
CRYSTAL OF 17α - HYDROXYPROGESTERONE

A.Szyczewski and R.Krzyminiewski

Institute of Physics, A.Mickiewicz University
Grunwaldzka 6, PL 60-780 Poznań, Poland

17α - Hydroxyprogesterone is a natural metabolite of the female hor-
mone, progesterone and differs from the latter only by the presence
of a hydroxyl group at C17 atom /Fig. 1/. Orthorhombic structure crystals [1] were grown from acetone solution and γ-irradiated with a
dose of 160 kGy. The kind and value of the hyperfine split, obtained
from EPR spectra, suggest that the radical is formed by hydrogen atom
abstraction from carbon atom C6. The unpaired electron is delocalized
on the four carbon atoms C6, C5, C4 and C3 and on the oxygen atom O3.
The mean values of the hyperfine interaction tensors are: A=1.13 mT,
A_1=1.96 mT, A_2=2.58 mT and the tensor of g coefficient /factor/:
g_x=2.0027, g_y=2.20037, g_z=2.055.
The above described type of radical is characteristic for nonhydrated
steroid single crystals such as progesterone and cholest-4-en-3-on
[2, 3] in which the hydrogen bond is not observed. The crystal of
17α -hydroxyprogesterone is also anhydrous but the hydrogen bond is
formed between OH /17/ and the oxygen atom O /3/ of the neighbouring
molecule. However, the presence of this bond does not influence the
type of radical formed.
Interesting conclusions may be drawn from a comparison of the above
results with the results of investigations of γ-irradiated single
crystals of hydrated testosterone [4]. Testosterone has the structure
of A and B rings of the steroid skeleton the same as 17α -hydroxypro-
gesterone. However, depending on the crystallographic structure the
radicals formed under the influence of γ-irradiation in testosterone
appear at the atoms of C_2 carbon /monoclinic form/ or O_3 oxygen
/rhombic form/. Thus, they are completely different from those formed
in 17α -hydroxyprogesterone.
Further studies are necessary in order to find out whether the differences are due to the presence of water in the crystals or to the
differences in the geometry of hydrogen bonds.

H. Fischer, H. Heimgartner (Eds.)
Organic Free Radicals
© Springer-Verlag Berlin Heidelberg 1988

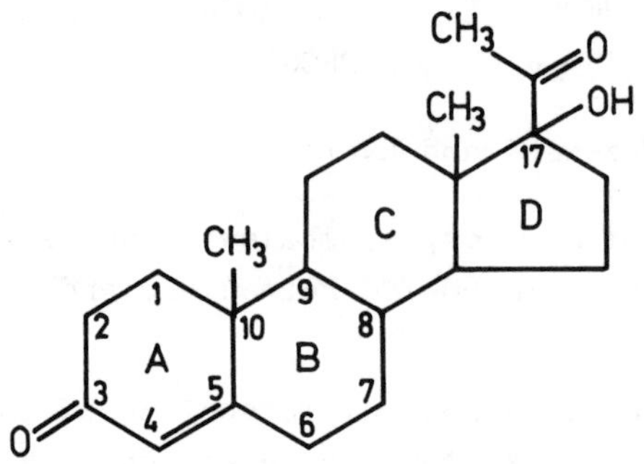

Fig. 1. The structure of 17∝ -hydroxyprogesterone molecule

References

1 Declercq J P, Germain G, Van Meeresche M /1972/ Cryst Structure
  Comm 1:9
2 Krzyminiewski R, Masiakowski J, Pietrzak J, Szyczewski A /1982/
  J Magn Resonance 46:300
3 Krzyminiewski R, Hafez AM, Pietrzak J, Szyczewski A /1983/ J Magn
  Resonance 51:308
4 Szyczewski A, Krzyminiewski R, Hafez AM, Pietrzak J /1986/ Inter
  J Radiat Biol 50:841

ENANTIOSELECTIVE ELECTRON TRANSFER DURING
THE FREE RADICAL CHAIN REDUCTION OF KETONES

D.D. Tanner and A. Kharrat

Department of Chemistry, The University of Alberta
Edmonton, Alberta, T6G 2G2, Canada

Abstract: Enantioselective electron transfer from a chiral di-
hydropyridyl radical to racemic mixtures of several bicyclic ketones
was demonstrated to be one of the propagation steps in the radical
chain reduction of these ketones. A transition state model
rationalizes the stereochemical outcome of these reductions.

INTRODUCTION

The reduction of a variety of ketones by a number of N,N'-substitut-
ed dihydronicotinamides (DHNA's) has recently been shown to proceed
via a free radical chain reduction mechanism whose propagation
sequence contains an electron transfer process (1-4). When a chiral
DHNA, 1, was used to reduce an enantiomeric mixture of ketones (±-
fenchone, 2) enantioselective reduction takes place to yield
optically active products (exo/endo fenchyl alcohols) and unreacted
enantiomerically enriched fenchone (3,4). When a mixture of
(4R,9R)-1 and (4S,9R)-1 was used for the reduction the same
stereochemical results were obtained.

DISCUSSION AND RESULTS

A number of other enantiomeric ketone pairs are reduced by a similar
radical chain process (see Table 1). Reduction of the ketones with
3, however, shows no enantioselectivity even though stereoselective
reduction of ketones having prochiral carbonyls gives stereoselec-

H. Fischer, H. Heimgartner (Eds.)
Organic Free Radicals
© Springer-Verlag Berlin Heidelberg 1988

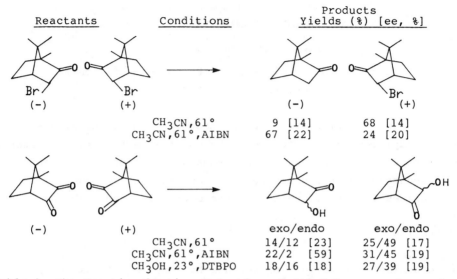

	Reactants	Conditions	Products Yields (%) [ee, %]	
	(-)	(+)	(-)	(+)
		CH₃CN,61°	9 [14]	68 [14]
		CH₃CN,61°,AIBN	67 [22]	24 [20]

(Table continued)

	(-)	(+)	exo/endo	exo/endo
		CH₃CN,61°	14/12 [23]	25/49 [17]
		CH₃CN,61°,AIBN	22/2 [59]	31/45 [19]
		CH₃OH,23°,DTBPO	18/16 [18]	27/39 [19]

Table 1. The Enantioselective Reduction of (±)-3-Bromocamphor and (±)-Camphoroquinone.

tive reduction with this DHNA. Reduction with (4R,9S)-**1** shows enantioselective reduction, but with the opposite antipodal selectivity. Transition state models for enantioselective electron transfer can be proposed which require: the amide carbonyl to be locked out of the plane of the ring, the medium sized methyl group of the chiral amide to be away from the amide carbonyl, and the ketone carbonyl to be situated with the smallest group away from the amide carbonyl with dipoles aligned, see Fig. 1.

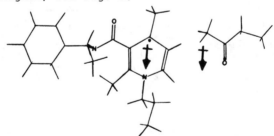

Fig. 1. Transition State Model for Enantioselective Electron Transfer.

REFERENCES

1 Tanner DD, Singh HK, Kharrat A, Stein AR (1987) J Org Chem 52:2142
2 Tanner DD, Kharrat A (1988) J Org Chem 53:1646
3 Tanner DD, Kharrat A (1988) Abstracts, 3rd Chemical Congress of North America, Toronto, Ontario, 0415
4 Tanner DD, Kharrat A (1988) J Am Chem Soc 110:0000, in press

HOMOSOLVOLYSIS

J. Smith and J. M. Tedder

Department of Chemistry, University of St. Andrews,
St. Andrews, Fife, KY16 9ST, Scotland.

We have studied the effect of substituents on the solvolytic reactions of some substituted bromoalkanes.   In particular we have studied the

homosolvolysis of 4,4'-disubstituted bromodiphenylmethane.   The most reactive compounds are those in which X is a donor group and Y is an acceptor group.

We have also studied the homosolvolysis of bromine substituted esters by measuring the disappearance of the nitroxide signal.   Compounds with an acceptor group and a donor group [e.g. $BrCH(OMe)CO_2Me$] were much more reactive than compounds with two donor or two acceptor groups [e.g. $BrCH(CO_2Et)_2$.

Singh H, Tedder JM (1980) J Chem Comm 1095
Smith J, Tedder JM (1988) J Chem Research (S) 1081

H. Fischer, H. Heimgartner (Eds.)
Organic Free Radicals
© Springer-Verlag Berlin Heidelberg 1988

HYDROGEN TRANSFER ACTIVATION
IN RADICAL TELOMERIZATION

A.B.Terent'ev,N.A.Kardanov,M.A.Moskalenko,
R.G.Petrova,T.D.Churkina,N.N.Godovikov

Institute of Organo-element compounds
Academie of Science USSR
II78I3 USSR,Moscow,Vavilov st.28

It was shown, that use of $Mn_2(CO)_{IO}$ enables to initiation radical reactions of organo-element compounds,that react with repture of ele ment-hydrogen bonds ($RSH,R_3SiH,(RO)_2P(O)H$),with olefins and to increase efficiently the efficiency of hydrgen transfer in telomerization.

Radical-complex activation of hydrogen transfer in addition and tel omerization is a perspective route to selective synthese of function al group containing organo-element compounds.

The reactions,that in the presence of peroxydes proceed as telomeri zation (with formation of higher telomers) in the case of $Mn_2(CO)_{IO}$ allow to produce adduct in good yeald, also with easily polymerizable monomers. Ethylene telomerization with $(C_2H_5)_3SiH$ in the presence of $Mn_2(CO)_{IO}$ results in adduct alone,whereas with peroxydes the part of adduct in sum of telomers is only 64%. For methylacrilate the yeald of $(C_2H_5)_3SiCH_2CH_2CO_2CH_3$ was 75%.

Thereaction of C_4H_9SH with acrilyc monomers (initiator-$Mn_2(CO)_{IO}$) re sulted in formation only adduct by relation monomer+mercaptane up to IO. We belive,that Mn-containing complex take part in these reactions

Radical mechanism of these processes was confirmed by ESR- it was identified intermediately forming radicals $(C_2H_5)_3\overset{\bullet}{S}i$ and $C_4H_9SCH_2\overset{\bullet}{C}H-CO_2CH_3$. I-hexene telomerisation with $(C_2H_5O)_2\overset{\bullet}{P}(O)H$ in the presence of $Mn_2(CO)_{IO}$ proceeds with significant induction time and termic post effect. Telomers $(C_2H_5O)_2P(O)(CH_2CHC_4H_9)_nH$ (T_n) distribution changed in favor of the first telomer (with perxydes $T_I/T_2=5$, with $Mn_2(co)_{IO}$ $T_I/T_2=I3$).

Relativ kinetic datum showed,that hydrogen transfer on the chain transfer step in this reaction in the presence of $Mn_2(CO)_{IO}$ is of 30-60 times more effective,than in usual peroxydes initiated process

H. Fischer, H. Heimgartner (Eds.)
Organic Free Radicals
© Springer-Verlag Berlin Heidelberg 1988

SELENIUM RADICAL CATIONS PROMOTED
FUNCTIONAL GROUP TRANSFORMATIONS

M. Tiecco, L. Testaferri, M. Tingoli, D. Bartoli, and D. Chianelli

Istituto di Chimica Organica, Università di Perugia
06100-Perugia, Italy

Abstract: Selenium radical cations are suggested to be the reactive intermediates in the oxydeselenenylation of aryl alkyl selenides promoted by oxidizing agents in hydroxylated solvents.

In recent years organic selenium reagents have become a very powerful tool in organic synthesis (1). Some free radical selenium reactions have also found useful synthetic applications (2). Selenium radical ions, on the contrary, have been scarcely investigated and their potential importance as reactive intermediates in organic synthesis is substantially unexplored. We have recently reported that selenium radical anions can be employed to effect the selective cleavage of the selenium alkyl bond in the alkoxy- and thioalkoxyaryl alkyl selenides and in vinyl alkyl selenides (3,4).

$$\text{Ar(XR)SeR} \xrightarrow{\ e\ } \text{Ar(XR)SeR}^{\cdot -} \longrightarrow \text{Ar(XR)Se}^{-} + \text{R}\cdot \qquad (X = O, S)$$

$$\text{ArSeR} \xrightarrow{-e} \text{ArSeR}^{\cdot +} \longrightarrow \text{ArSe}\cdot + \text{R}^{+}$$

We now report that selenium radical cations suffer a similar fragmentation to afford selenium radicals and alkyl carbocations. Very simple procedures to effect several functional group transformations, in which the formation and the fragmentation of selenium radical cations represent the crucial steps, have been developed. These reactions were simply carried out by oxidizing diphenyl diselenide in hydroxylated solvents in the presence of an unsaturated compound. As indicated in the following scheme these processes consist in the production of the phenylselenium cations, in the oxyselenenylation of the unsaturated compounds and in the oxydeselenenylation of the formed addition products. Thus, the entire process can be effected in one-pot. In the presence of an excess of the oxidizing agent the PhSe cations can be regenerated and in most cases the reaction requires only catalytic amounts of diphenyl diselenide. Besides persulphate anions other oxidizing agents can be employed with similar good results.

H. Fischer, H. Heimgartner (Eds.)
Organic Free Radicals
© Springer-Verlag Berlin Heidelberg 1988

$$PhSeSePh + S_2O_8^{--} \longrightarrow 2\ PhSe^+ + 2\ SO_4^{--}$$

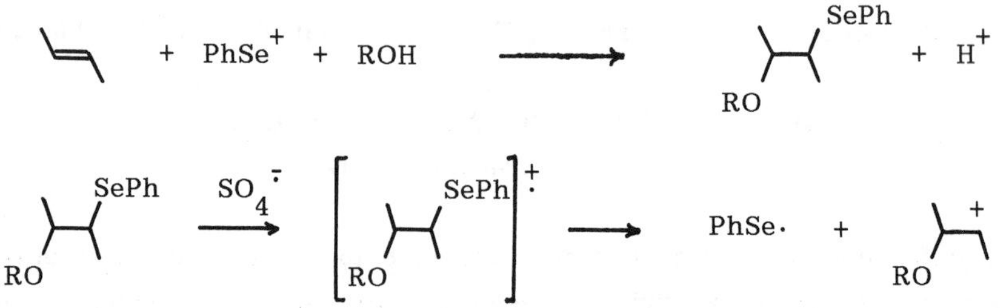

(scheme: alkene) $+ PhSe^+ + ROH \longrightarrow$ (product with SePh and RO) $+ H^+$

(scheme: RO–SePh compound) $\xrightarrow{SO_4^{\cdot-}}$ [RO–SePh radical cation] $^{\cdot+} \longrightarrow PhSe\cdot +$ (RO cation)

$$PhSe\cdot + SO_4^{\cdot-} \longrightarrow PhSe^+ + SO_4^{--}$$

The results obtained from these experiments are summarized in the scheme reported below. Alkenes **1** gave a mixture of 1,2-, **2** , and 1,1-dialkoxyalkanes **3** , whereas vinyl bromides **1** (R_1 = Br) gave good yields of the α-alkoxy acetals **2** (R_1 = OR). α-Keto acetals **5** were obtained from 1-alkynes **4** as well as from methyl ketones **6** ; non terminal alkynes **7** gave α-keto ketals **8** .

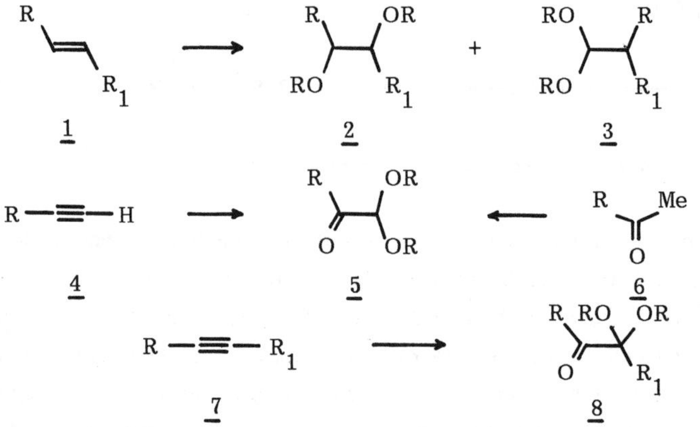

(Scheme showing structures 1, 2, 3, 4, 5, 6, 7, 8)

Several examples of these conversions will be presented and discussed.

References

1 Paulmier C (1986) Selenium Reagents and Intermediates in Organic Synthesis, Pergamon Press, Oxford

2 Back TG (1987) In: Liotta D (ed) Organoselenium Chemistry. John Wiley & Sons

3 Tiecco M, Testaferri L, Tingoli M, Chianelli D, Montanucci M (1983) J Org Chem 48: 4289

4 Testaferri L, Tiecco M, Tingoli M, Chianelli D (1986) Tetrahedron 42: 4577 and references cited therein

ALTERNATING REACTIVITY OF RADICALS COORDINATED TO CHELATED TRANSITION METALS OR METALLOENZYMES AND CHEMICAL CARCINOGENESIS

A.Tkáč and E. Hanušovská - Tkáčová [x]

Institute of Physical Chemistry, Slovak Technical University,
[x]Cancer Research Institute, Slovak Academy of Sciences
812 37 Bratislava, Czechoslovakia

Differences in the radical chemistry of normal and tumorous rat liver cells are indicated by ESR applied simultaneously with the Clark-oxygen membrane electrode. In cytosol of healthy cells the reactivity of primary radicals formed in the one-electron transfer step from chelated transition metals (Fe,Co,Cu,Cr) and metallo-enzymes to tert.butyl hydroperoxide (BuOOH) or to H_2O_2 is stepwice decreased in an H-transfer cascade closed by biological H-donors (tocopherol,ascorbic acid, glutathione).In this way healthy mitochondria are effectivey shielded against auto-oxidation and damage by secondary oxygen or nitrogen centred radicals generated from unhindered phenolic,aminoaromatic,aminoazoaromatic and polyaromatic carcino-genes[1,2].The opposite is truth for the rapidly proliferating Zajdela hepatoma as-citic tumor cells depleted successively from the metalloenzymes Mn-superoxide dis-mutase and catalase during the cell transformation.The original H_2O_2 decomposing capacity is renewed,when to the tumorous mitochondria (ZH-M) healthy ones (RL-M) are added (Table 1).The changed level of Mn-SOD is measured according to the in-tensity of the 6-line ESR signal of Mn(II) , manifested, when mitochondria are treated with ascorbic acid.Cytochrome c and cytochrome oxidase seam to be the main targets of ZH-M for electron abstraction by ran-domly formed radicals leading to irreversible oxidation of ferro- to ferry- hemoproteins (high-spin Fe^{III},g= 4.27 at 140 K, Table 2). Hemichromes still decompose hydroperoxides to coordinated peroxyl radicals already at physio-logical temperature (g= 2.0147),when the electron transfer proceeds in non-polar media and when stronger coordinating agents are excluded.The in-termediately formed free peroxyl radicals (g= 2.0092) from BuOOH in aqueous media of liver cells and mitochondria are detectable only after rapid freezing the reacting components to 140 K (TABLE 2, UQ⁻ =ubisemiquinone) .

INITIAL RATE OF O_2 EVOLUTION from 0.03 % H_2O_2 in 18 ml water suspension , T = 293 K	
5 mg protein	10^{-3} Ms⁻¹
Catalase	2.800
RL-M	0.460
RL-cells regenerating	0.170
ZH-M	0.016
ZH-ascites	0.006
ZH-M + RL-M	0.470
ZH-ascites + RL-cells	0.160

RL=rat liver,M=mitochondria
ZH= Zajdela hepatoma
TABLE 1.

The mean life-time and the operating radius of coordinated radicals at 300 K is a sensitive function of the oxidative- and spin-state of the transition metal,of the

H. Fischer, H. Heimgartner (Eds.)
Organic Free Radicals
© Springer-Verlag Berlin Heidelberg 1988

Table 2.

RAPID FREEZING to 140 K in 10s		10^{12} spins/30 mg proteins in 1 ml H_2O			
		NORMAL RAT LIVER		ZAJDELA.HEPATOMA	
ESR	g	CELLS	MITOCHONDRIA	ASCITES	MITOCHONDRIA
$UQ^{\cdot-}$	2.0040	2.4	2.7	2.1	2.8
Fe(III)4.27		0.5	0.0	1.5	22.2
ADDING 0.05 ml of BuOOH to 0.3 ml of colloidal solution					
$RO_2^{\cdot}$	2.0092	240.0	6.0	30.0	20.0
Fe(III) 4.27		3.0	4.0	5.0	120.0
HEATING THE SAMPLE TRANSIENTLY FOR 1 min to 300 K					
$RO_2^{\cdot}$	2.0092	0.0	0.0	0.0	0.0
Fe(III) 4.27		18.0	16.0	6.8	160.0

actual ligand sphere and local reactivity of the heterogenous biological environment[3].The restricted reactivity against H-transfer inactivation of coordinated radicals prepared from some H-donor carcinogens in non-polar media (benzene,toluene,CCl_4 paraffin oil,saturated fat acids,acetone) is in polar solvents (H_2O,ethylalcohol , ethylether,pyridine) or after decomplexation with stronger coordinating agents (e.g. bases of nucleic acids) cancelled and the original reactivity of the free radical is renewed.The alternating reactivity of coordinated radicals during their migration in hydrophobic cell compartments is suggested as a possible way of carrying radicals through defending barriers and membranes being not deactivated to the vicinity of the genetic information.

All components used in this study for generation of coordinated radicals at ambient temperature in non-polar,non-complexing media or of free radicals by rapid freezing technique in polar media can act as initiators or promotors of tumor growth : H_2O_2[4-6], t-BuOOH[7-9],transition metals[1,2,10,11],denaturated hemoproteins [3].

References :
1 Tkáč A (1987) Development in polymer stabilization-8,Scott G (ed).,pp.61-179,Elsevier London.
2 Tkáč.A,Bahna L (1983) Neoplasma 30:197
3 Tkáč A (1986) In:Schilov A (ed) Fundamental research in homogenous catalysis,pp.817-836,Gordon and Breach,London
4 Ito A, Naito M, Naito Y, Watanabe H (1982) Jap.J.Cancer Res.73:315
5 Florence TM (1983) Chemistry in Australia 50:116
6 Meneghini R, Hoffmann ME (1980) Biochim.Biophys.Acta 608:167
7 Pryor WA (1985) In:Fiwely JW and Schwass BE(eds)Xenobiotic Metabolism,Nutritional Effects,pp77-96,ASC Symposium Series No 277
8 Lwoff A (1953) Bact.Rev. 17:269
9 Taffe BG, Takahashi N, Kensler TW and Mason RP (1987) J.Biolog.Chem. 262:12143
10 Wilson RL (1972) Iron Metabolism,Ciba Foudation Sym. 51,New Ser.p.331,Elsevier, Amsterdam
11 Sano K, Assano T, Tanishima T, Sasaki T (1980) Neurological Res. 2:253

OXIDATION OF CROWDED POLYARYLPHOSPHINES AND DIPHOSPHINES.ESR AND ELECTROCHEMICAL STUDIES

P. TORDO

SREP CNRS UA 126 Université de Provence   13397 Marseille Cedex 13, France

Many tervalent phosphorus compounds are good electron donnors and for many years phosphoniumyl radicals, $L_3P^{+\cdot}$, have been suggested as intermediates in various chemical (1) and electrochemical (2) reactions involving tervalent phosphorus compounds. However these cationic species are very transient in solution which makes their ESR characterization particularly challenging. Thus the electrochemical oxidation of different phosphines L_3P, within the cavity of an ESR spectrometer (3), led only to the characterization of the corresponding dimeric cations, $(L_3PPL_3)^{+\cdot}$.

We have prepared a large series of triarylphosphines whose molecular geometry and steric strain were modified by changing the number and position of methyl groups on the aryl ligands. Then we investigated the influence of the molecular geometry on the anodic behavior of these phosphines and on the ESR features and lifetime of the corresponding cation radicals.

The oxidation potential was shown to depend strongly on the bending angle (α) at the phosphorus. Its value was reduced by 0.68 V, on going from the triphenylphosphine ($\alpha \approx 25°$) to the tridurylphosphine ($\alpha \approx 19°$). Within a series of triarylphosphines exhibiting aproximately the same bending angle, satisfactory Hammet correlation was found with $E_{1/2}$..

The steric effect of the methyl substituents was particularly striking when the two ortho positions of each phenyl group were substituted. Thus the peak current ratio I_p^a/I_p^c was found very close to one (at 0.1 V s^{-1}) for the trixylyl, trimesityl and triduryl phosphine while it dropped to zero for different isomers bearing only one ortho methyl substituent.

The triarylphposphines exhibiting a significantly reversible anodic oxidation at moderate potential sweep rates,

H. Fischer, H. Heimgartner (Eds.)
Organic Free Radicals
© Springer-Verlag Berlin Heidelberg 1988

yielded ESR signals when they were oxidized within the cavity of an ESR spectrometer. The principal feature of these isotropic ESR spectra comprises a doublet resulting from a relatively large phosphorus coupling (240 G)and these spectra can be obviously inferred to the phosphoniumyl radicals.

The existence of unresolved couplings with the methyl hydrogens was established by the synthesis of d_{27}-trimesityl phosphine exhibiting fully deuterated methyl groups, and the generation of the corresponding phosphoniumyl radical, whose linewidth was approximately reduced by half with respect to that of the undeuterated material.

Hybridization ratios calculated from the isotropic and anisotropic coupling constants indicated that these cation radicals retain a pyramidal geometry although the electron loss was accompanied by a substantial flattening of the pyramidal geometry of the starting phosphines.

With the strongly crowded phosphoniumyl radicals $L_3P+.$, (L = mesityl, duryl or o-xylyl) we never observed any evidence of the formation of dimeric cations $(L_3PPL_3)^{+.}$. On the other hand these crowded cation radicals were shown to be unreactive with water but afforded a very fast reaction with molecular oxygen.

Diphosphine cation radicals were detected during the anodic oxidation of crowded tetraaryldiphosphines. In contrast with the nitrogen analogs, the magnitude of the phosphorus coupling supported the preference of these radicals to adopt a pyramidal geometry. For the diphosphine cation radicals, the high P-P bond length makes the three electron stabilization negligible and far too small to balance the energy required to make the two phosphorus centers planar

1 POWELL RL, HALL CD (1969) J.Am.Chem.Soc., 91, 5403. KOTTMANN H, SKARZEWSKI J, EFFENBERGER H (1987) Synthesis, 9, 797.

2 KARGIN NYu, NIKITIN EV, PARAKIN OV, ROMANOV GV, PUDOVIK AN, (1978) Dokl.Akad.Nauk.SSSR, 242, 1108. EFFENBERGER F, KOTTMANN H, (1985) Tetrahedron, 19, 4171.

3 GARA WB, ROBERTS BP (1978) J.Chem.Soc.,Perkin Trans.2, 150.

ESR TECHNIQUE IN A 2 mm RANGE OF WAVELENGTHS AND ITS PHYSICO-CHEMICAL APPLICATIONS

Yu.D.Tsvetkov

Institute of Chemical Kinetics and Combustion,
Novosibirsk 630090, USSR

Some characteristic features of a 2 mm ESR spectroscopy as compared with a 3 cm ESR: the influence of high H_o on anisotropic hfi, on the properties of spectra in polycrystals, on line broadening due to spin-lattice relaxation are considered. A spectrometer developed for physico-chemical applications is described. Applications of a 2 mm ESR spectroscopy to divide the spectra of stable free radicals in mixture, to determine the magnetic resonance parameters, to study molecular mobility and dimerization are given. All investigations have been done on organic free radicals mainly nitroxide type in different glassy solutions. In conclusion the possible approaches to other chemical applications of a 2 mm ESR spectroscopy are considered.

H. Fischer, H. Heimgartner (Eds.)
Organic Free Radicals
© Springer-Verlag Berlin Heidelberg 1988

CYTOCHROME P450 CATALYZED CAGE RADICAL REACTIONS

V. Ullrich

Faculty of Biology,
University of Konstanz

Cytochrome P450 is now recognized as a group of hemoproteins with a thiolate ligand at the fifth coordination site of the heme [1]. The established and main function of these hemoproteins is the activation of molecular oxygen for hydroxylation reactions. Although many details of the monooxygenases are known, only indirect knowledge of the insertion mechanism of the oxygen atom into the substrates is available. From isotope effects, a radical nature of the intermediate activated oxygen species seems likely which would point to a hydrogen atom abstraction as the primary step. From comparison with related oxo species in other hemoproteins, one can conclude that the thiolate ligand may be responsible for the radicalic nature of the active oxygen complex. The second step required to form the hydroxyl or epoxy group could be either an oxidation of the carbon radical to a carbocation or alternatively a transfer of an OH-radical from the ferryl complex to the carbon radical.

We recently succeeded in the isolation of two isomerases which catalyse the rearrangement of prostaglandin, 9,11-endoperoxide to prostacyclin or thromboxane, respectively [3,4]. According to their spectral characteristics, both enzymes are cytochrome P450 proteins and therefore also must contain a thiolate ligand at the heme although no monooxygenase activity is associated with their physiological function. With iodosobenzene, however, thromboxane was able to transfer an oxygen atom to a CH-bond of a prostaglandin derivative which confirms the oxene transferase activity also of these enzymes [5]. Spectral studies revealed that the endoperoxide rearrangement is initiated by a coordination of one oxygen atom of the endoperoxide group to the ferric iron. Interestingly, prostacyclin synthase coordinates with the 11-oxygen whereas thromboxane synthase binds to the 19-oxygen atom. According to model systems and to experiments with different endoperoxide substrates, a homolytic cleavage of the endoperoxide occurs [6]. In prostacyclin synthase the

H. Fischer, H. Heimgartner (Eds.)
Organic Free Radicals
© Springer-Verlag Berlin Heidelberg 1988

free alkoxy radical at C9 adds to the 5 double bond and forms the five-membered heterocycle, followed by oxidation of the carbon radical by the iron-sulfur-moiety. In the case of thromboxane synthase, the 11-alkoxy radical isomerizes to a 10-keto-11-carbon radical intermediate. This radical must also be oxidized by the remaining iron-sulfur-moiety to a carbocation for which a concerted ring closure to the bicyclic thromboxane system is feasible. As a second product, malondialdehyde and a C^{17} acid also arises which can be explained by a free radical fragmentation process.

We conclude that the biosynthesis of prostacyclin and thromboxane proceeds via a cage radical process in which carbon radicals are oxidized by a ferric thiyl species. This is in analogy with the postulated hydroxylation mechanism via a carbocation pathway.

References

1 Ruf HH, Wende P, Ullrich V (1979) Characterisation of hemin mercaptide complexes by electronic and esr spectra
J Inorg Biochem 11: 189-204

2 Guengerich FP, Macdonald TL (1984) Accounts Chem Res 17:9-16

3 Ullrich V, Graf H (1984) Trends in Pharmacol Sci 5: 352-355

4 Haurand M, Ullrich V (1985) J Biol Chem 260: 15059-15067

5 Hecker M, Baader WJ, Weber P, Ullrich V (1987) Eur J Biochem 169:563-569

6 Hecker M (1988) Dissertation, Universität Konstanz

SINGLE-ELECTRON-TRANSFER (SET) LEADING TO RADICAL SPECIES IN THE REDOX REACTIONS OF CYCLIC PEROXIDES WITH BIOLOGICAL SUBSTRATES

W. Adam, L. Hasemann and F. Vargas

Institute of Organic Chemistry
University of Würzburg, am Hubland, D-8700 Würzburg, F.R.G.

B. Epe, D. Schiffmann and D. Wild

Institute of Toxicology
University of Würzburg, Versbacher Straße 9, D-8700 Würzburg, F.R.G.

In view of the genotoxicity [1] of 1,2-dioxetanes $\underline{1}$, which are efficient thermal sources of triplet excited carbonyl products

$\underline{1a}(R=CH_3)$
$\underline{1b}(R=CH_2OH)$
$\underline{2}$
$\underline{3}$

[2], it was of interest to establish whether the DNA damage was of photochemical character or the result of redox reactions of such cyclic peroxides with cell components. The living cell guards itself against such "oxidative stress" by deactivation of theh oxidants, e.g. the glutathione defense mechanism [3]. Indeed, the cyclic peroxides $\underline{1}$-$\underline{3}$ are efficiently reduced to their respective dihydroxy products by thiols, particularly glutathione and cysteine. With sulfides. e.g. thionine, oxygen transfer is observed, resulting in sulfoxides and the corresponding deoxygenated products of the cyclic peroxides [1]. As minor pathways, the 1,2-dioxetanes suffer catalytic decomposition [4], while C-H insertion products are formed by all three cyclic peroxides $\underline{1}$-$\underline{3}$ with thiomethyl substrates. As initial step for this transformation a single-electron-tranfer process is proposed [1], leading to a radical ion pair, from which all the observed products can readily be rationalized (eq 1). A variety of biologically

H. Fischer, H. Heimgartner (Eds.)
Organic Free Radicals
© Springer-Verlag Berlin Heidelberg 1988

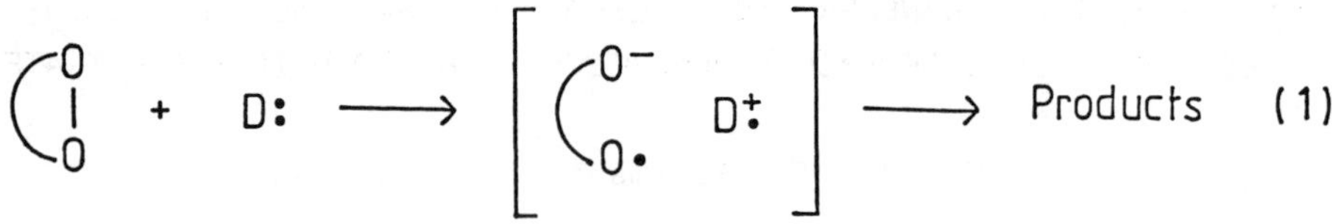

$$\mathrm{C}\!\!\begin{matrix}O\\ \|\\ O\end{matrix}\ +\ D\colon\ \longrightarrow\ \left[\ \mathrm{C}\!\!\begin{matrix}O^-\\ \\ O\cdot\end{matrix}\ \ D^{\underline{+}}\ \right]\ \longrightarrow\ \text{Products}\qquad(1)$$

significant substrates can serve as electron donors, including 2-mercapto-4-methylpyrimidine, phenothiazines, chloropromazine, ascorbic acid, tocopherol, ß-carotene, N-benzyl dihydronicotinamide, NADH, and aromatic amines such as tetramethylphenylenediamine, dimethyl-p-anisidine, tetramethyl-4,4'-diaminodiphenyl, 9-methylacridine, etc. For the phenothiazine the characteristic radical cation was detected, while efficient chemiluminescence was observed for aromatic amines with low oxidation potentials, presumably via the CIEEL mechanism [5]. The mechanistic details of this novel single-electron-transfer process and the biological implications will be discussed.

1 Adam W, Epe B, Schiffmann D, Vargas F, Wild D (1988) Angew Chem
  Int Ed Engl 27:429
2 Cilento G, Adam W (1988) Photochem Photobiol (in press)
3 Sies H (1986) Angew Chem 98:1061
4 Wilson T, Chia-Sen D (1973) Chemiluminescence and
  Bioluminescence, Cormier M J, Hercules D M, Lee J (eds), Plenum
  Press, New York, pp 265-283
5 Schuster G B (1979) Acc Chem Res 12:366

CHAIN TRANSFER MECHANISM IN THE RADICAL TELOMERIZATION INVOLVING THE BROMINE CONTAINING POLYHALOMETHANES

F.K.Velichko and T.T.Vasiljeva

Institute of Organo-element compounds Academie
of Science USSR

II78I3 USSR Moscow Vavilov str. 28

It is estimated a linear correlation between chain transfer constants $C_n = k_{tr}/k_p$ and electron affinity (E_A) of telogens in telomerization.

For the case of telomerization of vinyl chloride with $CHBr_3$ and CCl_nBr_{4-n} (n = 0-3) BPM) we have found a good correlation of C_n with E_A values of BPM (when $E_A > 0,4$ eV) which can be determined from transfer band of ferrocene-BPM charge transfer complexes (CTC)

$$C_I = (20,46\pm0,47)E_A - (9,48\pm0,39) \qquad r = 0,999 \ (I)$$
$$C_2 = (I06,80\pm I3,67)E_A - (45,25\pm II,32) \qquad r = 0,984 \ (2)$$

This correlation indicates that there is a possibility of CTC formation in the process of transfer Br atom from a telogen RBr to the growing telomer radical $RM_n^{\bullet}$ (M - monomer). We suppose that linear dependence of (I) (2) type takes place when the chain transfer step involves CTC formation between $RM_n^{\bullet}$ and BPM, which is succeeded by anion radical arising with the result that electron transfers from $RM_n^{\bullet}$ to BPM

$$\overset{\bullet}{C}Br_3 + n \ CH_2=CHCl \longrightarrow CBr_3CH_2CHCl)_n^{\bullet} \ (RM_n^{\bullet})$$
$$RM_n^{\bullet} + CBr_4 \longrightarrow (CTC \ RM_n^{\bullet} \ \overset{\bullet}{C}Br_4) \longrightarrow (RM_n^+ \ \overset{\bullet}{C}\bar{Br}_4) \longrightarrow (RM_n^+ \ Br^- \ \overset{\bullet}{C}Br_3)$$
$$\longrightarrow RM_nBr + \overset{\bullet}{C}Br_3$$

The correlation estimated allows to calculate C_n from spectral data and reveals a new approach to investigation of free radical halogen abstraction reactions.

H. Fischer, H. Heimgartner (Eds.)
Organic Free Radicals
© Springer-Verlag Berlin Heidelberg 1988

FREE RADICAL PRODUCTION FROM THE TRIPLET STATE OF NALIDIXIC ACID

G.Vermeersch*,J.C.Ronfard-Haret+,M.Bazin* and R.Santus*

* Lab.de Physique, Faculté de Pharmacie, F-59045 LILLE-Cedex, FRANCE
+ Lab.de Photochimie Solaire, C.N.R.S., rue H.Dunant, F-94320 THIAIS, FRANCE
* Lab de Physico-Chimie de l'Adaptation Biologique associé INSERM U.312,
Muséum d'Histoire Naturelle, F-75005 PARIS

The use of Nalidixic acid (NAH) as an antibacterial agent is limited by side effects (Ocular phototoxicity (1) and skin photosensitization (2)).We have thus investigated the primary photochemical processes of this molecule in presence or in absence of biological substrates using 355 nm Laser Flash Photolysis technique.

The deprotonated form (NA^-) which is predominently present at physiological pH (nalidixic acid pK_a = 6.2 (3)) has been studied.

RESULTS.

Triplet Properties.

The triplet-triplet transient absorption spectrum of (NA^-) is shown in figure 1-a (λmax = 620 nm, $\epsilon_T\Phi_T$ = 5700 $M^{-1}cm^{-1}$, 0.6 < Φ_T < 1). In absence of substrate the triplet decays following mixed first and second order (Triplet-triplet annihilation) kinetics with k_T = 1.5 10^4 s^{-1} and k_{TT} = 4.5 10^9 $M^{-1}s^{-1}$.

The transient NA^- triplet is quenched by oxygen (k_QO_2 = 3.2 10^9 $M^{-1}s^{-1}$) leading to singlet oxygen ($^1O_2$) (4) with Φ_Δ = 0.15 (5). The transient triplet is also efficiently quenched by the aromatic amino acids (AA) : Tryptophan (k_QTrp = 2.5 10^9 $M^{-1}s^{-1}$) and Tyrosine (k_QTyr = 1.4 10^9 $M^{-1}s^{-1}$) leading to the generation of the nalidixic acid radical anion ($NA^{\cdot-}$) and of radicals derived from AA (fig.1-b).

$$^3NA^{-*} + AA \xrightarrow{+OH^-} NA^{\cdot-} + AA^\cdot$$

Nalidixic Acid Radical Anion Properties.

Under anoxic conditions,the spectrum of the radical anion peaks at 650 nm (ϵ = 3000 $M^{-1}cm^{-1}$) (fig.1-b). It decays following second order kinetics (k = 5.1 10^9 $M^{-1}s^{-1}$)

The radical anion reacts efficiently with oxygen (k_QO_2 = 2.0 10^9 $M^{-1}s^{-1}$) leading to production of the superoxide radical ion ($O_2^{\cdot-}$). The presence of $O_2^{\cdot-}$ is evidenced by reduction of ferri-cytochrome C which is inhibited by SOD (Superoxide dismutase).

H. Fischer, H. Heimgartner (Eds.)
Organic Free Radicals
© Springer-Verlag Berlin Heidelberg 1988

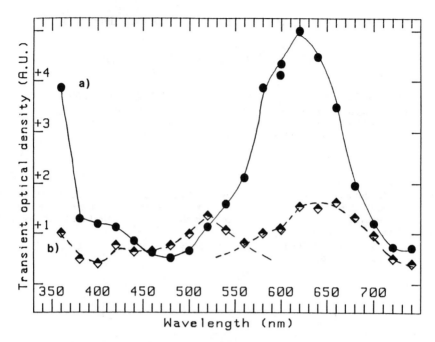

Figure 1 :Transient spectra of NA⁻ (4.3 10⁻⁴ M) + Trp (8.6 10⁻⁴ M) in buffered (pH = 9.2) water solution recorded

a) immediately after laser pulse (T-T absorption spectrum)

b) 4.3 μs after laser pulse, showing the decomposition into NA·⁻⁻ and Trp· spectra.

CONCLUSION

NA⁻ is not only a $^{1}O_2$ producer but its first excited triplet state also forms in a good yield radical species resulting from redox reactions in presence of oxidizable substrates.

Such free radical reactions may be involved in the phototoxic skin response accompanying nalidixic acid therapy.

References

1 -Marshall B.Y. (1967) The Practitioner, 199, 213-215
 -Fraunfelder F.T. (1982) Drug-induced ocular effects and drugs interactions, pp 33-35,Lea & Febiger, Philadelphia
2 -Zelickson A.S. (1964) J.A.M.A., 190, 556
  -Magnus I.A. (1976) Dermatological Photobiology, pp 213-215, Blackwell, Oxford
  -Brauner G.T. (1975) Am.J.Med., 58, 576-580
3 -Staroscik R. and Sulkowska J. (1971) Acta Pol.Pharm., 28, 601-606
4 -Moore D.E., Hemmens V.J.and Yip H. (1984) Photochem.Photobiol., 39, 57-61
5 -Dayhaw-Barker P. and Truscott T.G. (1988) Photochem.Photobiol., 47, 765-767

NEW PROCEDURES OF HOMOLYTIC ALKYLATION OF

HETEROAROMATIC BASES:SYNTHESIS OF C-NUCLEOSIDES

E.Vismara, F. Minisci and F. Fontana

Department of Chemistry, Polytechnic, Milan, Italy

C-Nucleosides of heteroaromatic bases have recently attracted a large interest for the biological activity in antibiotic and antitumoral fields[1].

We have recently developped a large variety of methods for the homolytic alkylation of heteroaromatic bases[2].

In order to have a selective process the radical source must have oxidizing character for the oxidation of the intermediate heteroaromatic radical adduct. This fact contrasts in several cases with sensibility of substrates such as sugars, towards oxidants agents.

A particularly useful source of alkyl radicals is provided by iodine abstraction from alkyl iodides, by aryl (eq.1) or methyl radicals (eq.2)

$$R\text{-}I \quad + \quad Ar^{\cdot} \quad \longrightarrow \quad R^{\cdot} \quad + \quad ArI \qquad k > 10^9 \ M^{-1} \ s^{-1} \quad (1)$$

$$R\text{-}I \quad + \quad Me^{\cdot} \quad \longrightarrow \quad R^{\cdot} \quad + \quad MeI \qquad k > 10^6 \ M^{-1} \ s^{-1} \quad (2)$$

However the most common sources of $Me^{\cdot}$ and $Ar^{\cdot}$ in an oxidizing system involve the primary formation of oxygen-centered radicals, which are very reactive towards C-H bond in α-positions to alcohols and ethers (always present in sugar molecules). Among the numerous radical sources developed by us for this purpose, two procedures proved to be suitable also with complex molecules as iodosugars. The first one involves the reaction of H_2O_2 with the iododerivative in DMSO and catalytic amount of Fe(II) salt (eq.3).

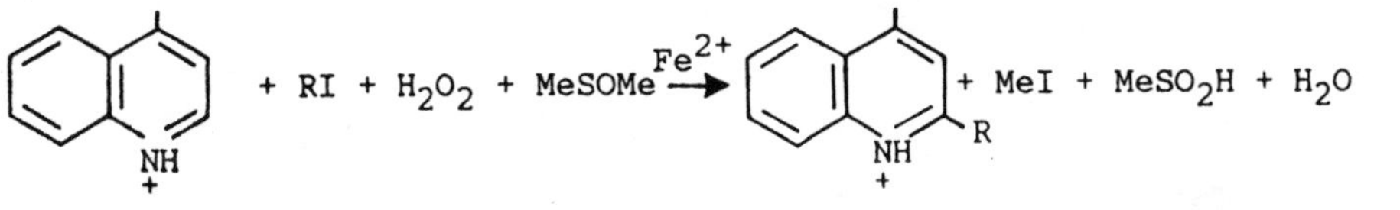

$$+ \ RI \ + \ H_2O_2 \ + \ MeSOMe \ \xrightarrow{Fe^{2+}} \ + \ MeI \ + \ MeSO_2H \ + \ H_2O \qquad (3)$$

H. Fischer, H. Heimgartner (Eds.)
Organic Free Radicals
© Springer-Verlag Berlin Heidelberg 1988

The alkyl radical is generated in this case from $^{\cdot}OH$ radical and DMSO (eq. 4)

$$HO^{\cdot} + MeSOMe \longrightarrow Me - \overset{\overset{\displaystyle \cdot O \quad OH}{\diagdown \diagup}}{S} - Me \longrightarrow Me^{\cdot} + MeSO_2H \quad (4)$$

The high reactivity and low selectivity of the OH radical with iodosugars is minimized by the excess of DMSO. On the contrary the radical reacts selectively with iodosugars in iodine abstraction and that makes possible the synthesis of C-Nucleosides.

A different procedure involves thermal (eq. 5) or redox decomposition (eq. 6) of diacetyl peroxide as source of the Me radical:

$$(MeCOO)_2 \longrightarrow 2 \ MeCOO^{\cdot} \longrightarrow 2 \ Me^{\cdot} + 2 \ CO_2 \quad (5)$$

$$(MeCOO)_2 + Fe^{2+} \longrightarrow Fe^{3+} + MeCOOH + MeCOO^{\cdot} \longrightarrow Me^{\cdot} + CO_2 \quad (6)$$

Also in this case the primary radical is oxigen-centered ($MeCOO^{\cdot}$) but its rate of decarboxylation is very high ($>10^9 \ M^{-1} \ s^{-1}$) making irrelevant the intermolecular, competitive processes.

References

1 Robins R.K., Revanvar G.R. (1985) Med. Res. Rev. 5:273

2 a) Fontana F., Minisci F., Vismara E. (1987) Tetrahedron Letters 28:6373

   b) Vismara E. (1983) La Chimica e l'Industria 65:34.

POLYMER-BOUND PYRYLIUM SALTS

M.Vondenhof, J.Mattay[+]

Institut für Organische Chemie, RWTH Aachen, Prof.Pirlet-Str.1

Pyrylium salts are effective sensitizers for various photoreactions[1]. If electronically excited, they act as electron
acceptors in photochemically induced electron transfer reactions. Two new pyrylium salts have been synthesized, both
being fixed to a polymeric backbone. One of these is prepared
in two simple steps, starting from Merrifield resin. They are
insoluble in organic solvents and thus provide examples for
heterogeneous electron transfer.

We have tested them[2] as sensitizers for the dimerization of
1,3-cyclohexadiene and phenyl vinyl ether. Other reactions
that were subjected to our research were the addition of phenyl
vinyl ether to 1,3-cyclohexadiene and the photooxygenation of
1,1-diphenylethylene.

We could distinguish from triplet reactions by utilizing the
long wavelength excitation band ($\lambda > 400$nm) of the pyrylium
system. The dimerization of 1,3-cyclohexadiene was catalyzed
with 16% yield in 7 hours, compared to 31.6% yield with the

H. Fischer, H. Heimgartner (Eds.)
Organic Free Radicals
© Springer-Verlag Berlin Heidelberg 1988

homogeneous, monomeric 2,4,6-triphenylpyrylium tetrafluoroborate in a comparable concentration. The difference is probably caused by the heterogeneous reaction conditions in the
first case. The dimerization of phenyl vinyl ether by sensi-
tization with our polymeric electron acceptors was as effective as the use of triplet sensitizers, whereas the high
yields of the two (cis and trans) products obtained by use of
homogeneous electron acceptors such as 2,4,6-triphenylpyrylium-
tetrafluoroborate, could not be reached.

With our new electron acceptors, we could achieve the addition
of phenyl vinyl ether to 1,3-cyclohexadiene, leading mainly to
the endo- and exo-Diels-Alder adducts, which is not possible
with triplet sensitizers. The cyclodimerization of 1,3-cyclo-
hexadiene and phenyl vinyl ether were both suppressed.

References:

1) a) J.Mattay, Angew.Chem.99(1987),849; Angew.Chem.Int.Ed.
      Engl. 26(1987),825
   b) J.Mattay, Nachr.Chem.Tech.Lab. 36(1988),April-Issue
2) J.Mattay, G.Trampe, M.Vondenhof, EPA Newslett., in press.

STABILISATION AND STEREODYNAMICS
OF FREE RADICALS

J. C. Walton

Department of Chemistry, University of St. Andrews, St. Andrews,
Fife, KY16 9ST, U.K.

K. U. Ingold

National Research Council of Canada, Division of Chemistry,
Ottawa, Canada, K1A OR6

D. C.Nonhebel

Department of Chemistry, University of Strathclyde,
Glasgow, G1 1XL, U.K.

Abstract: Stabilisation energies (SEs) of a number of free radicals
have been determined. Methods of predicting radical SEs are reviewed
and some chemical consequences will be examined.

The presence of thermodynamic stabilisation in a radical intermediate
is often a good indication that the route *via* this species will be
favoured. Polyenyl radicals have large SEs and recent determinations
of their SEs will be examined, together with empirical and semi-
-empirical correlations with structural properties of the radicals.
The EPR method of determining SEs from internal rotation barriers has
been applied to a number of radical types including C-D radicals.[1] A
test of the recent suggestion[2] that C-D radicals should be stabilised
in solvents of high dielectric constant will be described.

Rearrangements of radicals derived from bicyclo[n.1.o]alkanes (1),
bicyclo[n.2.o]alkanes (2), spiro[2.n]alkanes (3) and bicyclo[n.1.1]-
alkanes (4) have been investigated by EPR spectroscopy and product

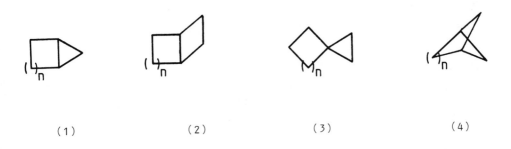

(1)   (2)   (3)   (4)

H. Fischer, H. Heimgartner (Eds.)
Organic Free Radicals
© Springer-Verlag Berlin Heidelberg 1988

analysis.   The effects of radical stabilisation, ring strain and stereoelectronic factors on the mode and rate of β-scission will be described.

References

1 Walton JC (1984) Rev Chem Intermed S 249
2 Katritzky AR, Zerner MC, Karelson MM (1986) J Am Chem Soc 108:7213

ELECTROCHEMISTRY OF FREE RADICALS: SUBSTITUENT EFFECTS ON THE REDOX POTENTIALS OF BENZYL RADICALS

D.D.M. Wayner,[*] B.A. Sim, and D. Griller

Division of Chemistry, National Research Council of Canada
Ottawa, Ontario, Canada   K1A 0R6

Abstract: The electrochemical oxidation and reduction potentials of a number of 3-, and 4-substituted benzyl radicals have been measured using the photomodulation voltammetry technique.

INTRODUCTION

Recently, we have reported a method, photomodulation voltammetry, PMV, for measuring directly the electrochemical oxidation and reduction potentials of free radicals in solution (1-3). This technique uses modulated photolysis (in this case a 1000 W Hg/Xe arc lamp whose output is modulated with a mechanical chopper) to generate the radicals thus causing the concentration of the radicals to modulate at the same frequency. The use of phase sensitive electrochemical detection allows us to discriminate in favour of those processes which occur at the modulation frequency. This method of detection leads to an increase in signal-to-noise of several orders of magnitude and radicals with lifetimes of 10^{-3} s and concentrations of ca. 10^{-7} M can be detected easily.

RESULTS AND DISCUSSION

All of the benzyl radicals were generated by photolysis of di-tert-butyl peroxide (DTBP) followed by hydrogen abstraction from the corresponding toluene by the tert-butoxy radical. The oxidation potentials correlate well with σ^+ giving a slope of ca. 550 mV and a correlation coefficient of 0.98 (Figure 1a). Similarly, the reduction potentials correlate with σ^- giving a slope of ca. 770 mV and a correlation coefficient of 0.98 (Figure 1b). The correlation coefficients are not significantly improved when an extended Hammett treatment, including $\sigma \cdot$ (4), is applied.

Although there has been interest in the electrochemical properties of radicals and ions (see references cited in ref. 3), there have

H. Fischer, H. Heimgartner (Eds.)
Organic Free Radicals
© Springer-Verlag Berlin Heidelberg 1988

not been, to our knowledge, any systematic studies reported on the redox potentials of benzylic radicals and ions. We have found that the reduction potentials are more sentitive to substitution than the oxidation potentials. Since the bond energies of the parent hydrocarbons are not expected to change by more than 1-2 kcal mol^{-1} (5), the changes in the oxidation and reduction potentials can be related, albeit crudely, to the relative pK_R's and pK_a's of the corresponding carbocations and carbanions, respectively (an additional assumption, of course, is that the measured potentials are reversible). In this case, the pK_R's and pK_a's cover a range of 7.5 and 15 pK units respectively for the substituents measured.

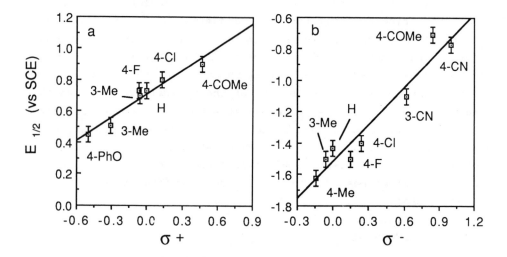

Figure 1. (a) oxidation potentials (vs. SCE) vs. σ^+ and, (b) reduction potentials (vs. SCE) vs. σ^- of substituted benzyl radicals measured in acetonitrile/DTBP (9:1), 0.1 M tetrabutylammonium perchlorate.

Acknowledgements: We gratefully acknowledge financial support from the President's Fund of the National Research Council of Canada.

References:

1 Wayner DDM, Griller D (1985) J Am Chem Soc 107:7764
2 Wayner DDM, Dannenberg JJ, Griller D (1986) Chem Phys Lett 131:189
3 Wayner DDM, McPhee DJ, Griller D (1988) J Am Chem Soc 110:132
4 Wayner DDM, Arnold DR (1985) Can J Chem 63:2378
5 Pryor WA, Church DF, Tang FY, Tang RH (1980) In: Pryor WA (ed)
  Frontiers of Free Radical Chemistry. Academic Press, New York

DISSOCIATION AND ION—MOLECULE REACTIONS
OF ALKANE RADICAL CATIONS.
TIME-RESOLVED FDMR [1]

D. W. Werst, L. T. Percy and A. D. Trifunac

Chemistry Division
Argonne National Laboratory, Argonne, IL 60439 USA

Alkane and olefin radical cations, fundamental reaction intermediates in organic chemistry and in the radiation chemistry of hydrocarbons, have been characterized by time-resolved Fluorescence Detected Magnetic Resonance (FDMR) in pulse radiolysis of alkane liquids and solids from 30K to room temperature. Studies of olefin cations in neat and mixed alkane liquids have elucidated the mechanism for conversion of alkane cations into olefin cations by loss of H_2. Alkane cations are extremely short-lived in neat alkane liquids, and their detection and characterization in pure hydrocarbon systems by magnetic resonance techniques have been elusive. Stabilization of alkane cations can result from (1) lowered temperature and/or (2) isolation of the cation from neutral parent molecules (as in alkane mixtures). Our observations suggest that alkane cations rapidly undergo proton transfer to neutral alkane molecules, in competition with geminate recombination.

1Work performed under the auspices of the Office of Basic Energy Sciences, Division of Chemical Science, US-DOE under contract number W-31-109-ENG-38

H. Fischer, H. Heimgartner (Eds.)
Organic Free Radicals
© Springer-Verlag Berlin Heidelberg 1988

ENDOR OF METALLOPROTEINS AT
X-BAND AND Q-BAND MICROWAVE
FREQUENCIES

M. Werst and B. M. Hoffman

Department of Chemistry
Northwestern University, Evanston, IL   60208   USA

One of the common difficulties in the ENDOR studies of many metalloproteins is that the resonances of non-protonic elements in the site of interest frequently are obscured by the more intense proton ENDOR signal. This makes it difficult and often impossible to detect, assign and analyze the ENDOR response from other nuclei. The proton pattern is centered at the proton Larmor frequency, ν_H, which is proportional to the microwave frequency; at X-band and g~2, ν_H~14 MHz. The overlap problem can be eliminated by performing the ENDOR measurements at Q-band (35 GHz). At Q-band, g~2 corresponds to a magnetic field of 12500 G and the proton pattern is shifted to a center frequency of ν_H~50 MHz. In contrast, the center-frequency of the ^{14}N, ^{17}O and ^{57}Fe, etc. resonances is determined by the hyperfine coupling constant, and they remain at 0-30 MHz. Q-band measurements have been made on the copper sites of cytochrome oxidase from different organisms, the type I copper sites of copper blue proteins and the iron sulphur cluster of reduced aconitase. These studies, which have revealed distinct signals due to ^{14}N and ^{57}Fe nuclei not previously seen at X-band, will be discussed.

H. Fischer, H. Heimgartner (Eds.)
Organic Free Radicals
© Springer-Verlag Berlin Heidelberg 1988

WHY BROMINE IS USED
IN THE AUTOXIDATION OF P-XYLENE
TO TEREPHTHALIC ACID

D.A. Young and W. Partenheimer

Amoco Chemical Company
P.O. Box 400, Naperville, IL 60566 U.S.A.

ABSTRACT

The oxygenation of p-xylene with dioxygen in the presence of a soluble cobalt (II) salt in an acetic acid solvent is the predominate commercial method to produce terephthalic acid. The addition of bromide effectively changes the radical mechanism(s) of the metals, eliminates the thermal barrier associated with cobalt only oxidations, and dramatically increases the rate, selectivity and yield of autoxidation.

BACKGROUND

Amoco Chemical Company is the world's largest producer of terephthalic acid. Approximately 97% of the terephthalic acid is reacted with ethylene glycol to make polyethylene terephthalate, this polyester is used for fibers, films, bottle resins, and various engineering plastics. In addition to terephthalic acid, Amoco uses its proprietary Mid-Century (MC) oxidation technology (soluble forms of cobalt (II), manganese (II), and bromide salts) to produce isophthalic acid, trimellitic acid, and a number of "Fine" specialty aromatic acids such as 2,6-naphthalene dicarboxylic acid, trimesic acid, and 5-tert-butylisophthalic acid.

MECHANISMS AND LIMITATIONS OF COBALT ONLY OXIDATIONS

The most widely accepted mechanism of the cobalt catalyzed oxidation of an alkylaromatic feedstock is via an electron transfer mechanism of cobalt (III) acetate. A simplified example of the reaction mechanism to the first isolatable nonradical product, the benzyl alcohol, is shown below:

$$CH_3C_6H_4CH_3 \ + \ Co \ (III) \ ===> \ CH_3C_6H_4CH_2 \cdot \ + \ Co \ (II) \ + \ H^+ \qquad (1)$$

$$CH_3C_6H_4CH_2 \cdot \ + \ O_2 \ ===> \ CH_3C_6H_4CH_2OO \cdot \qquad (2)$$

H. Fischer, H. Heimgartner (Eds.)
Organic Free Radicals
© Springer-Verlag Berlin Heidelberg 1988

$$CH_3C_6H_4CH_2OO\cdot \ + \ CH_3C_6H_4CH_3 \ ===> \ CH_3C_6H_4CH_2OOH + CH_3C_6H_4CH_2\cdot \qquad (3)$$
$$CH_3C_6H_4CH_2OOH + \ Co\ (II) \quad ===> \ CH_3C_6H_4CH_2O\cdot \ + Co\ (III) + OH^- \qquad (4)$$
$$CH_3C_6H_4CH_2O\cdot \ + \ CH_3C_6H_4CH_3 \ ===> \ CH_3C_6H_4CH_2OH + CH_3C_6H_4CH_2\cdot \qquad (5)$$

The alcohol can enter the chain sequence to form the benzaldehyde, and so on to the final product, the aromatic carboxylic acid.

Cobalt only catalyzed oxidations produce low yields of terephthalic acid (~15%) because of the electron-withdrawing nature of the carboxylic acid group (from the intermediate p-toluic acid). Decarboxylation of acetic acid and p-toluic acid by cobalt (III) also contributes to the low yields of terephthalic acid.

Increasing the reaction temperature fails to improve the yield in these oxidations, since the rate of cobalt (III) catalyzed decomposition of acetic acid increases with temperature and becomes the predominate reaction ~150°C.

WHY ADDITION OF BROMIDE

Alternative methodologies have been developed to overcome the unsatisfactory yields of terephthalic acid, including 1) partial oxidation, esterification, co-oxidation technique (Witten Process), 2) changing the cobalt structure and co-oxidation with aliphatic aldehydes and ketones (Eastman Process), and 3) the addition of bromide to the catalyst package (Amoco MC Process). Adding bromide to a cobalt only catalyzed oxidation effectively changes the oxidation from an electron transfer mechanism to a ligand transfer mechanism. Experimental verification is provided by examining the cobalt (III) concentration during an autoxidation. The cobalt (III) concentration is usually between 30 and 50% of the total cobalt concentration in a cobalt only oxidation as compared to 3% in a cobalt/bromide oxidation. We propose that the reduction in cobalt (III) concentration is caused by the rapid intramolecular electron transfer of a transitory cobalt (III) complex to a cobalt (II) - bromine atom complex. This complex can either dissociate to give a bromine atom or react directly with the aromatic feedstock. Manipulating the cobalt (III) mechanisms virtually eliminates the temperature barrier, increases selectivity and rate at a given temperature, and improves the yield of terephthalic acid to 90%. Further incentives include the utilization of higher temperatures, lower catalyst concentration, fewer by-products, and shorter residence times.

Author Index

Subject Index

E. Roduner

The Positive Muon as a Probe in Free Radical Chemistry

Potential and Limitations of the μ SR Techniques

1988. 8 figures, 7 tables. VII, 104 pages. (Lecture Notes in Chemistry, Volume 49). ISBN 3-540-50021-9

Contents: Introduction. – Experiments employing muons. – Theory. – The cyclohexadienyl radical. – Substituent effects on hyperfine coupling constants. – The process of radical formation. – Distribution of muons in substituted benzenes. – Radical reactions. – Summary and review.

The muon is introduced as a unique probe in radical chemistry, and the experimental technique is demonstrated to have considerable potential for further developments. Employing this method, extraordinary structural and kinetic effects have been encountered.
This book relates to muonated organic free radicals, the adducts of muonium to unsaturated molecules. The main aspects treated are:

- the experimental techniques and their sensitivity compared to conventional magnetic resonance methods,
- the applicability of the muon as a spin probe to measure absolute rate constants of fast radical reactions,
- the nature of the subnanosecond radiation chemical processes which occur upon injection of energetic muons in matter near the end of their thermalization tracks,
- geometric and electronic structure of the radicals formed, and isotope effects revealed by comparison with their normal hydrogenated analogues.

Springer-Verlag
Berlin Heidelberg New York
London Paris Tokyo Hong Kong

Springer

L. Eberson

Electron Transfer Reactions in Organic Chemistry

1987. 65 figures. XIII, 234 pages. (Reactivity and Structure, Volume 25). ISBN 3-540-17599-7

Contents: Electron Transfer, a Neglected Concept in Organic Chemistry. – Concepts and Definitions. – Theories of Electron Transfer in Organic Chemistry. – How to Use the Marcus Theory. – Experimental Diagnosis. – Reaction Between Organic and Inorganic Non-metallic Species. – Reaction Between Organic and Metal Ion Species. – Electron Transfer Reactions Between Organic Species. – Electricity and Light Promoted ET. – Electron Transfer Catalyzed Reactions. – ET and Polar Mechanisms; How Are They Connected? – Applications of Outer-sphere Electron Transfer. – Epilogue. – Subject Index.

The subject of the books is electron transfer reactions in organic chemistry, with the emphasis on mechanistic aspects. The theoretical framework is that of the Marcus theory, well-known from its extensive use in inorganic chemistry. The book deals with definitions of electron transfer, theory of electron transfer reactions (Marcus' and Pross-Shaik's approach) experimental diagnosis of electron transfer reactions, examples from inorganic/organic reactants and purely organic reactants, electro- and photochemical electron transfer, electron transfer catalyzed reactions, connections between electron transfer and polar mechanisms, and applications of electron transfer, such as electrosynthesis of organic chemicals, photochemical energy storage, conducting organic materials and chemiluminescence. The approach is new in so far as no comparable book has been published. The book will be of value to anyone interested in keeping track of developments in physical organic chemistry.

Springer-Verlag
Berlin Heidelberg New York
London Paris Tokyo Hong Kong

Springer